Lecture Notes in Mathematics

Edited by A. Dold and B. Eckmann

914

Max L. Warshauer

# The Witt Group of Degree k Maps and Asymmetric Inner Product Spaces

Springer-Verlag
Berlin Heidelberg New York 1982

**Author**

Max L. Warshauer
Department of Mathematics, Southwest Texas State University
San Marcos, TX 78666, USA

AMS Subject Classifications (1980): 10 C 05

ISBN 3-540-11201-4 Springer-Verlag Berlin Heidelberg New York
ISBN 0-387-11201-4 Springer-Verlag New York Heidelberg Berlin



Printing and binding: Beltz Offsetdruck, Hemsbach/Bergstr.
2141/3140-543210

TABLE OF CONTENTS

## INTRODUCTION

In these notes our goal has been to develop the algebraic machinery for the study of the Witt group $W(k,I)$ of degree $k$ mapping structures and the Witt group of asymmetric inner product spaces. We are particularly interested in their relationship which arises in an exact octagon which is studied both for a field $F$ and for the integers $Z$. We show that this octagon is the appropriate generalization of the Scharlau transfer sequence [Lm 201].

We have tried to develop the properties of these Witt groups in a self-contained and complete manner in order to make this accessible to a larger audience. When references are given they generally are specific including page numbers. However, we should mention two general references. First, The Algebraic Theory of Quadratic Forms by T.Y. Lam [Lm] develops the Witt group over a field of characteristic not equal to 2. Second, Notes on the Witt Classification of Hermitian Innerproduct Spaces over a Ring of Algebraic Integers by P.E. Conner [C] discusses Hermitian forms and the Witt group over an algebraic number field and ring of integers therein. Together these should provide any background material the reader might need.

Although the viewpoint we take is entirely algebraic, we would be remiss if we did not mention its topological motivation. Much of this work originated in our efforts to explain and exposit the important work of N.W. Stoltzfus, Unravelling the Integral Knot Concordance Group [Sf-1]. We shall describe this topological connection momentarily. First however we need to describe the algebraic objects at hand.

Our object is to define and study a Witt group $W(k,I)$ of degree $k$ maps, where $k$ is an integer, $D$ is the underlying ring, and $I$ is a D-module. This group consists of Witt equivalence classes of triples $(M,B,\ell)$ satisfying:

(1) $B: M \times M \to I$ is an I-valued inner product defined on the D-module $M$.

(2) $\ell: M \to M$ is a D-module homomorphism of $M$ satisfying $B(\ell x,\ell y) = kB(x,y)$. We refer to $\ell$ as a map of degree $k$.

A triple $(M,B,\ell)$ is metabolic (Witt equivalent to zero) if there is an $\ell$-invariant submodule $N \subset M$ with $N = N^{\perp}$. Here $N^{\perp}$ is the orthogonal complement. This enables us to define the Witt equivalence relation $\sim$ by:

$$(M,B,\ell) \sim (M_1,B_1,\ell_1) \text{ iff } (M \oplus M_1, B \oplus -B_1, \ell \oplus \ell_1)$$

is metabolic. The Witt equivalence class of $(M,B,\ell)$ is denoted $[M,B,\ell]$. The operation of direct sum $\oplus$ makes this collection of equivalence classes into a group $W(k,I)$. The identity is the Witt equivalence class of metabolic triples.

We also develop the basic properties of asymmetric inner product spaces $(M,B)$ where no symmetry requirement is placed on the inner product $B$. The key to understanding these asymmetric inner products is a "symmetry operator" $s$ satisfying $B(x,y) = B(y,sx)$

for all $x,y \in M$. As above we define the notion of metabolic by $(M,B)$ is metabolic if there is an s-invariant subspace $N \subset M$ with $N = N^{\perp}$. This leads us to define Witt equivalence as before, and there results the asymmetric Witt group.

There is a very interesting relationship between this asymmetric Witt group and the Witt group of degree $k$ maps. This comes from the squaring map $S: W(k,D) \to W(k^2,D)$ given by $[M,B,\ell] \to [M,B,\ell^2]$. For $D = F$ a field an eight term exact sequence is developed from $S$. This octagon involves the Witt group of asymmetric inner products $A(F)$ just described.

As a special case of the octagon we obtain the transfer sequence of Scharlau, Elman and Lam [Lm 201] and [E,L 23-25]. This appears as an exact octagon in which several terms vanish. We are able to reinterpret the kernel and cokernel of this transfer sequence. Thus, these Witt group $W(k,F)$ and $A(F)$ are an appropriate generalization of the classical Witt group $W(F)$, at least in so far as relating to and explaining the Scharlau transfer sequence.

We should like to prove exactness in the octagon over $Z$. In order to continue we have a boundary sequence which relates $W(k,Z)$ to $W(k,Q)$

$$0 \to W(k,Z) \to W(k,Q) \xrightarrow{\partial} W(k,Q/Z)$$

When $k = \pm 1$ this sequence is short exact. Using this we are able to prove exactness in the octagon over $Z$ when $k = \pm 1$ by comparing the octagon to the octagon over $Q$.

As we have said, in these notes we are developing the algebraic machinery to study the Witt group of degree $k$ maps and the asymmetric Witt group. We should discuss the topological motivation for this work. When $k = +1$ the Witt group $W(+1,Z)$ is the crucial element determining the bordism groups $\Delta_n$ of orientation preserving diffeomorphisms of $n$-dimensional closed oriented smooth manifolds. Medrano introduced the appropriate Witt invariant as follows.
Let $f: M^{2n} \to M^{2n}$ be an orientation preserving diffeomorphism of a closed, oriented, smooth $2n$-dimensional manifold $M^{2n}$. We consider then the pair $(M^{2n},f)$ in $\Delta_{2n}$. The degree $+1$ mapping triple associated to this pair is

$$(H^n (M^{2n};Z)/_{\text{torsion}}, B, f^*)$$

where $B(x,y) = \varepsilon_* ((x \cup y) \cap [M^{2n}])$, $\varepsilon_*$ is augmentation, $[M^{2n}]$ is the fundamental class, $\cup$ is cup product, $\cap$ is cap product, $f^*$ is the induced map on cohomology. If $(M^{2n},f)$ bounds, this triple is metabolic. Thus there is an induced homomorphism

$$I: \Delta_{2n} \to W(+1,Z).$$

The task of computing the bordism groups $\Delta_{2n}$, was completed by Kreck [K] who showed that this Witt invariant was essentially the only invariant for bordism of diffeomorphisms.

More generally, given a closed oriented $2n$-dimensional manifold together with a map $\ell$ of degree $k$, the corresponding Witt triple $(M,B,\ell)$ as above satisfies $B(\ell x,\ell y) = kB(x,y)$. We are thus led to examine the Witt group $W(k,Z)$. The Witt group of asymmetric inner product spaces arises in Quinn's work [Q] on open book decomposition. The relation of this to the Witt group arising in Kreck's work above is discussed by Stoltzfus in The Algebraic Relationship Between Quinn's Invariant of Open Book Decomposition and the Isometric Structure of the Monodromy [Sf-2]. In this he gives a geometric application of the exact octagon obtained in Chapter X.

The exact octagon (renamed "the eight fold way") has also been extended to the setting of the surgery obstruction groups by A.Ranicki, L. Taylor, and B. Williams. For a further discussion of the use of quadratic forms and the Witt group in topology the reader is referred to Alexander, Conner, Hamrick, Odd Order Group Actions and Witt Classification of Inner Products [ACH], and Stoltzfus Unravelling the Integral Knot Concordance Group [Sf-1].

We now describe briefly the organization of these notes. Chapter I lays out the inner product spaces and Witt groups we will be studying. We continue our study of the Wiit group in Chapter II by describing Witt invariants which will be used to compute the Witt group in many cases. These invariants include rank mod 2, discriminant, and signatures.

In Chapter III we study the characteristic and minimal polynomial of the degree $k$ map $\ell$ in a degree $k$ mapping structure $(M,B,\ell)$. This study is used in Chapter IV where we compute the Witt group $W(k,F)$ for $F$ a field. This is done by decomposing $W(k,F)$ according to the characteristic polynomial of $\ell$, and $A(F)$ according to the characteristic polynomial of $s$.

In Chapter V we develop an 8 term exact octagon which relates $W(k,F)$ to $A(F)$.

$$\begin{array}{ccccccccc}
 & \nearrow & W^{+1}(k,F) & \xrightarrow{S} & W^{+1}(k^2,F) & \rightarrow & W^{+1}(-k,F) & \searrow & \\
A(F) & & & & & & & & A(F) \\
 & \nwarrow & W^{-1}(-k,F) & \leftarrow & W^{-1}(k^2,F) & \xleftarrow{S} & W^{-1}(k,F) & \swarrow &
\end{array}$$

$S$ is the squaring map $[M,B,\ell] \to [M,B,\ell^2]$. We prove again the Scharlau, Elman, Lam transfer sequence, and see that the exact octagon we develop is its appropriate generalization. In order to study this octagon over $Z$, we relate $W(k,Z)$ to $W(k,Q)$ by a boundary sequence in Chapter VI.

Now we place an additional requirement on the degree $k$ mapping structure $[M,B,\ell]$ , namely:

(*) $\ell$ satisfies the monic integral irreducible polynomial $f(t)$.

The resulting Witt group of triples satisfying the additional requirement (*) is denoted $W(k,Z;f)$.

The action of $\ell$ induces a $Z[t]/(f(t))$ - module structure on $M$. To simplify the notation let $S = Z[t]/(f(t))$. Note that $S$ is only an order in the Dedekind ring of integers $O(E)$ of the algebraic number field $E = Q[t]/(f(t))$. This order $S$ may not be the maximal order $O(E)$.

The first step to understanding $W(k,Z;f)$ or $W(k,Z;S)$ (the same thing) is to study the group $W(k,Z;D)$ for $D = O(E)$ the maximal order. This group consists of Witt equivalence classes of inner product spaces $(M,B)$ in which $M$ is a finitely generated torsion free $D$-module. This is in contrast to $W(k,Z;S)$ in which we only insist that the module structure of $M$ lifts to the order $S = Z[t]/(f(t))$.

In Chapter VI we are interested only in $W(k,Z;D)$ and the resultant boundary sequence for the maximal order. We read this boundary sequence on the Hermitian level, where the  - involution on  E  is given by  $t \to kt^{-1}$  and  $t^{-1} \to k^{-1}t$.  One uses the following commutative diagram.

$$0 \to H(\Delta^{-1}(D/Z)) \to H(E) \xrightarrow{\partial(D)} H(E/\Delta^{-1}(D/Z))$$
$$\downarrow t \qquad\qquad \downarrow t \qquad\qquad \downarrow t$$
$$0 \to W(k,Z;D) \to W(k,Q;D) \xrightarrow{\partial(D)} W(k,Q/Z;D)$$

$\Delta^{-1}(D/Z)$  denotes the inverse different of  D  over  Z.  The vertical isomorphisms denoted by  t  are induced by the trace of  E  over  Q.

Thus the method employed for computing  $W(k,Z;D)$  is to study the corresponding boundary sequence in the isomorphic Hermitian case.  The image of  $\partial(D)$  is the group  $H(E/\Delta^{-1}(D/Z))$  which is computed as follows:

$$H(E/\Delta^{-1}(D/Z)) \xrightarrow{t} W(k,Q/Z;D) \xrightarrow{g} \bigoplus_{P = \overline{P}} W(k,F_p;D/P) \xleftarrow{tr} \bigoplus_{P = \overline{P}} \Pi(D/P)$$

Here we sum over all  - invariant maximal ideals  P  in  D.  The isomorphism  g  is induced by selecting a generator  $1/p$  for the p-torsion  in  Q/Z.  The trace map on finite fields from  D/P  to  $F_p$  induces the isomorphism  tr  with the Hermitian groups  $\bigoplus_{P = \overline{P}} H(D/P)$.

We use the letter  M  to denote  - invariant maximal ideals in S.  In order to study  $W(k,Z;S)$  one must use the following commutative diagram:

$$0 \to H(\Delta^{-1}(D/Z)) \to H(E) \xrightarrow{\partial(D)} H(E/\Delta^{-1}(D/Z))$$
$$\downarrow t \qquad\qquad \downarrow t \qquad\qquad \downarrow t \qquad\qquad\qquad\qquad \searrow tr$$
$$0 \to W(k,Z;D) \to W(k,Q;D) \xrightarrow{\partial(D)} W(k,Q/Z;D) \xrightarrow{g} \bigoplus W(k,F_p;D/P) \xleftarrow{tr} \bigoplus H(D/P)$$
$$\downarrow f_1 \qquad\qquad \downarrow f_2 \qquad\qquad \downarrow f_3 \qquad\qquad\qquad g \qquad\qquad\qquad tr \quad \downarrow tr$$
$$0 \to W(k,Z;S) \to W(k,Q;S) \xrightarrow{\partial(S)} W(k,Q/Z;S) \xrightarrow{g} \bigoplus W(k,F_p;S/M) \xleftarrow{tr} \bigoplus H(S/M)$$

First one computes $W(k,Z;D)$ for the maximal order by going to the Hermitian level and reading $\partial(D)$ in the group $\bigoplus_{P=\overline{P}} H(D/P)$. Then one forgets the D-module structure and remembers only the S-module structure via the maps $f_i$ to gain a computation for $W(k,Z;S)$.

Thus in Chapter VII we study non-maximal orders. Let us be explicit here in describing the key problems involved.

At every prime $P$ in $D$ there exists a canonically defined element $\rho(P)$ in $E/\Delta^{-1}(D/Z)$ with the following properties.

(1) The map of $O(E) \to E/\Delta^{-1}(D/Z)$ given by $\lambda \to \lambda\rho(P)$ induces an embedding of the residue field.

(2) We also have the map $Z \to Q/Z$ given by $n \to n/p$ which induces an embedding of $F_p = Z/pZ \to Q/Z$.

The element $\rho(P)$ is canonical in the sense that it makes the following diagram commute.

$$\begin{array}{ccc} O(E)/P & \to & E/\Delta^{-1}(D/Z) \\ \downarrow tr & & \downarrow t \\ F_p & \to & Q/Z \end{array}$$

The horizontal maps were just described. tr again denotes the map induced by the finite field trace. t denotes the map induced by the number field trace.

Thus we see that it is precisely these elements $\rho(P)$ which determine the isomorphism $tr^{-1} \circ g \circ t$ identifying $H(E/\Delta^{-1}(D/Z))$ with $\bigoplus_{P = \overline{P}} H(D/P)$. If we wish to use the commutative diagram just discussed to compute $W(k,Z;S)$, we must therefore study those elements $\rho(P)$. For it is in terms of these elements that one reads the local boundary

$$\partial(D,P): H(E) \to H(E/\Delta^{-1}(D/Z)) \to \bigoplus_{P = \overline{P}} H(D/P) \to H(D/P).$$

in such a way as to make our diagram commute. The last map is projection to the $P^{th}$ coordinate.

The first two sections of Chapter VII are due to Conner. In these we present his theorems which develop the fundamental properties of these canonical localizers $\rho(P)$. We complete our study of $W(k,Z;S)$ for non-maximal orders by discussing the finite field trace tr, and the maps $f_i$.

In Chapter 8 we finish our discussion of the boundary homomorphism. This includes the notion of coupling from Stoltzfus [Sf-1] between various $\partial(D)$, and a proof that the boundary

$$\partial: W(k,Q) \to W(k,Q/Z) \text{ is onto when } k = \pm 1.$$

In Chapter IX the terms and maps in the octagon are studied in detail. This, together with the information about the boundary map enables us to prove exactness in the octagon over $Z$ in Chapter X.

The idea to study this problem and the possibilities inherent in the program we have undertaken comes from Professor P.E. Conner. It is thus a pleasure to thank him for his help in this project without which the notes would ner have been written. The author feels fortunate to have had the opportunity to study under Professor Conner, whom we thank not only for his invaluable ideas, but also for his patience and understanding.

Further, for many of the ideas herein, he should receive credit.

The author is also grateful to Professor Stoltzfus for numerous conversations throughout this project. We also thank Professor Dan Shapiro at Ohio State University for his help and suggestions, not only on this paper but also when the author was just beginning to study mathematics. We also express our appreciation to Professors Cordes and Butts at Louisiana State University; to Professor A. Liulevicius at the University of Chicago; and to Professor A. Ross at Ohio State University for their interest in the author at various stages of his career.

Finally, the author wishes to express his gratitude to his wife, Hiroko, and parents Dr. and Mrs. Albert Warshauer who have given their constant support and encouragement.

## CONVENTIONS

A complete list of symbols and notations, as well as an index, can be found in the back. We number theorems, propositions, and definitions consecutively in each chapter. We refer to a theorem in the same chapter as it is numbered. However, when referring to a theorem from another chapter, we use a Roman numeral to indicate the chapter from which the theorem is taken.

The end of a proof is designated by the symbol □ . Occasionally, this symbol is also used alone, without a proof, to indicate that certain Lemmas or Propositions follow in a straightforward manner from the preceding, or that the proof is not difficult.

References are usually given together with a page number, eg. [Lm 201] refers to page 201 of reference [Lm].

Chapter I THE WITT RING

In this chapter we define the algebraic structures which are to be studied. Section 1 begins by describing the setting for our inner product spaces, and includes a brief discussion of prime ideals and valuations. In Section 2 we develop the elementary properties of these inner product spaces. Section 3 shows how to construct new inner product spaces out of old. In particular the operations of direct sum and tensor product are discussed. These operations later become addition and multiplication in the Witt ring.

Since our inner products need not be symmetric we are led to examine a symmetry operator in Section 4. An inner product space comprises part of the data of a degree k mapping structure which is defined in Section 5. A Witt equivalence relation is then defined on these structures.

Section 6 is concerned with selecting from each Witt equivalence class a certain "anisotropic" representative. We show that in certain cases this representative is unique.

## 1. Setting and notation

We are interested in developing a Witt classification for triples $(M,B,\ell)$ where $B: M \times M \to Z$ is a Z - valued inner product and $\ell: M \to M$ is a map of degree k. This means $B(\ell x, \ell y) = kB(x,y)$ for all $x, y \in M$. In order to accomplish this one must investigate algebraic number fields and the Dedekind ring of integers in these number fields. Further, we must carefully study the role of non-maximal orders in the Dedekind ring of integers. We begin by describing the setting.

Let D be a Dedekind domain [O'M 52] together with an involution - . The quotient field of D is E. The involution - extends to a Galois automorphism of E and we denote the fixed field of - by F. It may happen that F = E; in fact this is precisely the case when - is the trivial involution.

We shall also use the symbol O(E) to denote the Dedekind ring of integers D in E. Of course $O(F) = O(E) \cap F$. Let S be an order in D. This means S is a subring of D, containing 1, with the same rank as D [B 88]. We shall be particularly interested in principal orders S of the form $S = Z[t,t^{-1}]/(f(t)) = Z[\theta]$, where f(t) is a monic, integral irreducible polynomial. Thus S is an order in O(E), where E is the algebraic number field Q[t]/(f(t)). Although our primary interest is in these principal orders in algebraic number fields, the theory we shall develop applies to arbitrary orders in a Dedekind domain.

Let $\mathcal{P}$ be a prime ideal in O(E) = D. Then $\mathcal{P} \cap O(F) = P$ will denote the corresponding prime ideal in O(F) and $\mathcal{P} \cap S = \mathcal{M}$ will be the corresponding prime ideal in S. The image of $\mathcal{P}$ under the involution - is denoted by $\bar{\mathcal{P}}$. $\bar{\mathcal{P}}$ is also prime, hence maximal since D is a Dedekind domain [O'M 49]. Let I be a - invariant fractional ideal in D. Since I factors uniquely into a product of prime ideals, $I = \prod_i P_i^{n_i}$, $n_i \in Z$, it follows that $\bar{I} \subset I$ if and only if $\bar{I} = I$.

Associated to a prime P in O(F) or $\mathcal{P}$ in O(E) is a discrete non-Archimidean valuation $|\ |_P$, respectively $|\ |_{\mathcal{P}}$ [O'M 1]. $\mathcal{P}$ lies over P, meaning $\mathcal{P} \cap O(F) = P$ if and only if $|\ |_{\mathcal{P}}$ extends $|\ |_P$.

Now $|\ |_{\mathcal{P}}$ is a homomorphism from the units in E, E*, onto a cyclic, multiplicative subgroup of $R^+$. We may thus form $O_F(P)$ and $O_E(\mathcal{P})$, the local rings of integers associated to the primes P and

$P$. We have

$$O_E(P) = \{w \in E: |w|_P \leq 1\}.$$

We also have $O(E) = \bigcap_P O_E(P)$.

In the local ring of integers $O_E(P)$, associated to a prime $P$ is a maximal ideal

$$m(P) = \{ w \in E: |w|_P < 1\}.$$

In fact $m(P)$ is a principal ideal generated by some element $\pi \in O_E(P)$. Any such $\pi$ generating $m(P)$ is called a local uniformizer. Two such $\pi$ clearly have as quotient a local unit.

It is also useful for what follows to think of the exponential version of $|\ |_P$. Following [B,S 23] we denote this by $v_P$. $v_P : E^* \to Z$ is given by:

$$v_P(x) = n \text{ means } |x|_P = |\pi|_P^{\,n}.$$

We now have $m(P) = \{w \in E: v_P(w) > 0\}$. If $y \in F^*$, $v_P(y)$ is the exponent to which the prime ideal $P$ is raised in the factorization of $yO(E)$.

We shall review prime ideals and algebraic number theory results in Chapter II. Our object here has only been to establish notation for the inner product spaces which we are now ready to describe.

## 2. Inner products

Again $D = O(E)$ is a Dedekind domain. We consider pairs M,D which satisfy either:

(a) M is a finitely generated torsion free D-module (and hence projective since D is a Dedekind domain [R-1 85]) with K = I a - invariant fractional ideal in D or

(b) M is a finitely generated torsion D-module with K = E/I where E is the quotient field of D and I is a - invariant fractional ideal in D.

We are interested in studying the D-module $Hom_D(M,K)$, where M,K satisfy either (a) or (b) above. The D-module structure of $Hom_D(M,K)$ is now given by defining:

$$df(x) = f(\bar{d}x) \quad \text{for } d \in D,\ x \in M,\ f \in Hom_D(M,K).$$

Definition 2.1 A K-valued inner product space (M,B) over D is a finitely generated D-module M, together with a non-singular bi-additive mapping

$$B: M \times M \to K$$

satisfying $B(dx,y) = B(x,\bar{d}y) = dB(x,y)$ for all $x,y \in M$, $d \in D$.

B is linear in the first variable, conjugate linear in the second variable. Again, M and K are assumed to satisfy one of the standard assumptions (a) or (b) above.

It is still necessary to say what it means for B to be non-singular.

Definition 2.2 The map $B: M \times M \to K$ is non-singular

<u>provided the adjoint map</u> $Ad_R B$: $M \to Hom_D(M,K)$ <u>is a D-module isomorphism. By</u> $Ad_R B$ <u>we are denoting the right adjoint map, namely</u> $Ad_R B(x) = B(-,x)$.

The left adjoint map, $Ad_L B$ is similarly defined by $Ad_L B(x) = \overline{B(x,-)}$. We must conjugate in order to have $Ad_L B(x) \in Hom_D(M,K)$, ie. to make $Ad_L B(x)$ D-linear.

We have left out any symmetry requirements on B. This is taken care of by:

<u>Definition 2.3</u> <u>An inner product space</u> (M,B) <u>is</u> u <u>Hermitian provided</u> B <u>satisfies</u> $B(x,y) = u\overline{B(y,x)}$ <u>for all</u> $x,y \in M$, u <u>fixed</u> $u \in D$.

Since $B(x,y) = u\bar{u}B(x,y)$, it follows that $u\bar{u} = 1$ and u is a unit in D of norm 1 [S 60]. We see that 1 Hermitian is the usual notion of Hermitian, while -1 Hermitian corresponds to skew-Hermitian.

When the involution - is trivial, meaning the identity, 1 Hermitian is symmetric since $B(x,y) = B(y,x)$. Similarly, we define skew-symmetric, and u symmetric in the case that - is the identity.

Let us now return to study the module M in case (a), namely M is a finitely generated projective D-module.

We form the vector space $M \otimes_D E = V$ over E.

<u>Definition 2.4</u> <u>The rank of a finitely generated torsion free</u> D-<u>module</u> M <u>is the dimension of the vector space</u> $M \otimes_D E$ <u>over</u> E, E <u>being the quotient field of</u> D.

Thus viewed M is a D-lattice in V [O'M 209]. Hence, M splits as a direct sum $M = \bigoplus_{i=1}^{n} A_i$, $n$ = rank M, where each $A_i$ is a fractional ideal in D. In fact, [O'M 212], there is the splitting:

$$M = A_1 z_1 \oplus Dz_2 \oplus \ldots \oplus Dz_n$$

where $A_1$ = fractional ideal in D;

$$A_2 = A_3 = \ldots = A_n = D.$$

$\{z_i\}$ is a basis for V.

Since M splits as a sum of fractional ideals, and $\mathrm{Hom}_D$ is additive over direct sums, we are reduced to studying $\mathrm{Hom}_D(A,I)$, where A is a fractional ideal in D.

__Lemma 2.5__ *Let A, I be fractional ideals in D, with I a - invariant fractional ideal. Then the map*

$$\tau : \bar{A}^{-1} I \to \mathrm{Hom}_D(A,I)$$

*given by*

$$x \to \tau(x) \quad \text{with} \quad \tau(x)(c) = c\bar{x}$$

*is a D-module isomorphism. Here the D-module structure on $\mathrm{Hom}_D(A,I)$ is as previously defined.*

__Proof:__ First observe that $\tau(x) \in \mathrm{Hom}_D(A,I)$. To see this, note that $\tau(x)$ is clearly D-linear, and $\tau$ is a D-module homomorphism since:

$$\tau(dx)(c) = c(\overline{dx}) = (c\bar{d})(\bar{x}) = \tau(x)(\bar{d}c) = (d\tau(x))(c).$$

We must show that $\tau$ is an isomorphism.

(a) $\tau$ is 1-1: Suppose $\tau(x) = \tau(y)$. Then $c\bar{x} = c\bar{y}$, for all $c \in A$. Let $c \neq 0$, and cancel to obtain $\bar{x} = \bar{y}$, hence $x = y$.

(b) $\tau$ is onto: Let $f \in \mathrm{Hom}_D(A,I)$. We must show $f \in$ image $\tau$. Tensoring with E, we extend f to $\mathrm{Hom}_D(E,E)$, where E is the quotient field of D. Since the Lemma is clearly true for E, it follows that $f(c) = \bar{x}_0 c$, for $\bar{x}_0 \in E$. But $\bar{x}_0 c \in I$ for all $c \in A$. Hence $\bar{x}_0 \in A^{-1}I$, and $x_0 \in \bar{A}^{-1}\bar{I} = \bar{A}^{-1}I$. Therefore, $f(c) = \tau(x_0)$, and $f \in$ image $\tau$. □

<u>Theorem 2.6</u> *Let M be a finitely generated torsion free D-module, and I a - invariant maximal ideal in D. Then there is a canonical D-module isomorphism*

$$\phi: \ M \to \mathrm{Hom}_D(\mathrm{Hom}_D(M,I)I))$$

*given by*

$$\phi(x)(f) = \overline{f(x)} \ .$$

<u>Proof</u>: Recall again the module structure on $\gamma \in \mathrm{Hom}_D(\mathrm{Hom}_D(M,I),I))$ is given by: $(d\gamma)(f) = \gamma(\bar{d}f)$ , where $(\bar{d}f)(x) = f(dx)$. By the remarks immediately preceding Lemma 2.5, it suffices to prove the theorem for $M = A$ a fractional ideal.

We apply Lemma 2 .5 twice to obtain an isomorphism $\phi$ .

$\phi$: $A = \overline{(\bar{A}^{-1}I)I^{-1}} \xrightarrow{\tau} \mathrm{Hom}_D(\bar{A}^{-1}I,I) \rightarrow \mathrm{Hom}_D(\mathrm{Hom}_D(A,I),I))$. $\phi$ is given by the composition. We have then:

$$\begin{aligned}\phi(x)(f) &= \tau(x)(m) \quad \text{where} \quad \tau(m) = f, \text{ so } f(x) = x\,\bar{m} \\ &= \bar{x}\,m \\ &= \overline{f(x)}\end{aligned}$$

as claimed. We again observe that $\phi$ is a D-module isomorphism since:

$$\phi(dx)(f) = \overline{f(dx)} = \overline{(\bar{d}f)(x)} = \phi(x)(\bar{d}f) = ((d\phi)(x))(f) \ . \quad \square$$

We now wish to establish this result in case (b), namely when M is a finitely generated torsion D-module, with $K = E/I$, and $I = \bar{I}$ as usual. In order to do this we simplify matters by recalling the decomposition of finitely generated torsion modules over a Dedekind domain. This is done as follows. Let $P$ be a prime ideal in D, and let $M(P)$ denote the localization of M at $P$, so $M(P)$ is a torsion module over the principal ideal domain (p.i.d.) $D(P)$.

$$M(P) = M \text{ localized at } P = \{x \in M\colon P^n x = 0 \text{ for some } n\} \ .$$

We first decompose M as $M \simeq \bigoplus_P M(P)$. Then using the structure theorem for modules over a p.i.d., M is isomorphic to a direct sum of cyclic modules with each cyclic module of the form $D/P^i$ some $P$, $i \in Z$. Since $\mathrm{Hom}_D$ is additive over direct sums, without loss of generality we may assume M is a cyclic module, $M = D/P^i$.

In order to study $\mathrm{Hom}_D(M,E/I)$, we begin by simplifying E/I. Suppose the fractional ideal I factors as

$$I = P_1^{i_1} \cdots P_k^{i_k}\ , \text{ where the } P_i \text{ are maximal ideals in D.}$$

Let $\pi_i$ be a uniformizer for $P_i$. Then in the notation above, $I(P) = I$ localized at $P = IS^{-1}$ where $S = D - P$. Since $I = P_1^{i_1} \cdots P_k^{i_k}$, it is clear then that

$$I(P_j) = \pi^{i_j} D(P).$$

We can now simplify K = E/I.

<u>Lemma 2.7</u> $E/I = \bigoplus_P E/I(P)$.

<u>Proof</u>: Define $f\colon E/I \to \bigoplus_P E/I(P)$ by $e + I \to \bigoplus_P (e + I(P))$. f is clearly well-defined, and a homomorphism.

(a) f is 1-1: Suppose $f(e + I) = 0$. Then $e \in I(P)$ for all $P$. Thus $e \in I$, by [O'M 46]. Hence, f is 1-1.

(b) f is onto: Consider $\bigoplus_i (a_i + I(P_i)) \in \bigoplus_P E/I(P)$. We now apply the Strong Approximation Theorem, [O'M 42]. Letting $|\ |_i$ denote the $P_i$-adic valuation we can find $x \in E$ with $|x - a_i|_i < |\ I(P_i)\ |_i$, at the finite set of i when (a) $a_i \notin I(P_i)$ or (b) $I(P_i) \neq D(P_i)$; with $|x|_i \leq 1$ otherwise. It follows that $f(x + I) = \bigoplus_i (a_i + I(P_i))$, and f is onto. □

Notice that the only summand of E/I with $P$-torsion is $E/I(P)$.

Thus, when $M = D/P^i$ , we may apply Lemma 2.7 to obtain an isomorphism:

$$\mathrm{Hom}_D(M,E/I) \simeq \mathrm{Hom}_D(M,E/I(P)).$$

We further identify $E/I(P) \simeq E/\pi^j D(P)$. Here $I(P) = \pi^j D(P)$, where $\pi$ is a uniformizer for $P$ as before. Multiplication by $\pi^{-j}$ then gives the isomorphism $E/\pi^j D(P) \simeq E/D(P)$

The module structure of M lifts to D(P), and we claim there is a D(P)-module isomorphism $\tau$: $M(P) \to \mathrm{Hom}_{D(P)}(M(P),E/D(P))$. In order to define $\tau$, it suffices to consider the case $M(P) \simeq D/P^i$, since M(P) is a direct sum of cyclic modules. Let $x \in M$ be a generator for $D/P^i$. Then define $\tau(x)$ by:

$$\tau(x)(cx) = c/\pi^i \text{ where } c \in D(P),\ x \in M(P).$$

As in Lemma 2.5, $\tau(dx)(cx) = \bar{d}\, c/\pi^i = (d\tau)(x)(cx)$ defines $\tau$ over M. $\tau$ is clearly an isomorphism. Combining these isomorphisms over all cyclic module summands of M, we obtain:

<u>Lemma 2.8</u> <u>Let</u> M <u>be a finitely generated torsion</u> D-<u>module</u>. <u>Then there is a canonical isomorphism</u> $M \simeq \mathrm{Hom}_D(M,E/I)$. □

Exactly as Theorem 2.6 was proved, we now have:

<u>Theorem 2.9</u> <u>Let</u> M <u>be a finitely generated torsion</u> D-<u>module</u>, I <u>a</u> - <u>invariant fractional ideal</u>. <u>Then there is a canonical</u> D-<u>module isomorphism</u>:

$$\phi: \quad M \to \mathrm{Hom}_D(\mathrm{Hom}_D(M, E/I), E/I)$$

*given by*

$$\phi(x)(f) = \overline{f(x)} \; . \quad \square$$

For the purposes of the next section, we also need the following propositions which describe the relationship of $\otimes$ to Hom .

*Proposition 2.10* *There is a canonical isomorphism* $\psi$: $\mathrm{Hom}_D(A, \mathrm{Hom}_D(B, C) \to \mathrm{Hom}_D(A \otimes B, C)$, *where* A, B, C *are* D-*modules*.

*Proof*: [R-1 25] Define $\psi$ by $(\psi f)(a \otimes b) = f(a)(b)$. The inverse of $\psi$ is given by $(\psi^{-1} g)(a)(b) = g(a \otimes b)$ . $\square$

*Proposition 2.11* *Let* $X_1, X_2, Y_1, Y_2$ *be finitely generated projective* D-*modules*. *Then*

$$\mathrm{Hom}_D(X_1, Y_1) \otimes \mathrm{Hom}_D(X_2, Y_2) \approx \mathrm{Hom}_D(X_1 \otimes X_2 \; , \; Y_1 \otimes Y_2).$$

*The isomorphism is given by*

$$f \otimes g \to f\,g \; , \; \textit{where } (f\,g)(x_1 \otimes x_2) = f(x_1) \otimes g(x_2),$$

*and* f g *is extended to* $X_1 \otimes X_2$ *bilinearly*.

*Proof*: For $X_i$, $Y_j$ free, the assertion is clear using bases.

Now, if $X_i$, $Y_j$ are projective, then each is a direct summand of a free. Hom and $\otimes$ are additive over finite direct sums. Hence the isomorphism for free splits into isomorphisms for the summands. □

## 3. Constructing new inner products out of old

### 3.1 Direct Sums

Let $(M,B)$ and $(M_1,B_1)$ be two K-valued inner product spaces. The easiest way to construct a new inner product space is to form the sum.

$$(M,B) \oplus (M_1,B_1) \rightarrow (M \oplus M_1,\ B \oplus B_1)$$

Here $(B \oplus B_1)((x,y),(z,w)) = B(x,z) + B_1(y,w)$. It is clear that $B \oplus B_1$ is an inner product since the adjoint map

$$Ad_R(B \oplus B_1) : M \oplus M_1 \rightarrow Hom_D(M \oplus M_1,K)$$

splits as $Ad_R B \oplus Ad_R B_1$.

### 3.2 Tensor Products

The next operation on an inner product space is $\otimes$. Let $(M,B)$ and $(M_1,B_1)$ be two type (a) inner product spaces. In other words, $M$ and $M_1$ are both finitely generated projective D-modules. Assume that:

$$B: M \times M \rightarrow Y_1 \text{ and } B_1: M_1 \times M_1 \rightarrow Y_2 .$$

We have the adjoint isomorphisms:

$$Ad_R B: \quad M \to Hom_D(M,Y_1) \text{ and } Ad_R B_1: \quad M_1 \to Hom_D(M_1,Y_2).$$

Taking the tensor product of these, we obtain by Proposition 1.11

$$\begin{aligned} &Ad_R(B \otimes B_1) = \\ &Ad_R B \otimes Ad_R B_1: \quad M \otimes M_1 \to Hom_D(M,Y_1) \otimes Hom_D(M_1,Y_2) \\ &\qquad\qquad \simeq Hom_D(M \otimes M_1, \; Y_1 \otimes Y_2). \end{aligned}$$

This shows that the adjoint of $B \otimes B_1$ is an isomorphism, and hence $(M \otimes M_1, \; B \otimes B_1)$ is a $Y_1 \otimes Y_2$ - valued inner product space. We can identify $Y_1 \otimes Y_2$ with the product of ideals, $Y_1 Y_2$ .

3.3 Scaling an inner product

There is the operation of scaling an inner product.

Let (M,B) be a K-valued inner product space with $d \in E^*$. $E^*$ denotes the units in E, ie. $E^* = E - \{0\}$, since E is a field.

Clearly, (M,dB) is a dK-valued inner product space, where $(dB)(x,y) = d\,B(x,y)$. We may view this as a special case of tensor product, namely

$$(M,B) \otimes (D,B_d) = (M,dB), \text{ where } B_d(x,y) = dxy.$$

3.4 Tensoring with the quotient field of D

Given an inner product space (M,B), with M of type (a), we can form $(M,B) \otimes_D E = (M \otimes_D E, \; B \otimes 1)$, where we now denote the extension of B to the quotient field by $B \otimes 1$. When there is no confusion, we will write $B \otimes 1 = B$.

3.5 The discriminant inner product space

Let (M,B) be a D-valued inner product space, and suppose $M \simeq A \oplus D \oplus \ldots \oplus D$ with n factors, where n is the rank of M. We form the $n^{\underline{th}}$ exterior power, $\Lambda^n M \simeq A$, with inner product

$$\Lambda^n B: \ \Lambda^n M \times \Lambda^n M \to D \text{ defined by}$$

$$\Lambda^n B(x_1\Lambda\ldots\Lambda x_n, y_1\Lambda\ldots\Lambda y_n) = \text{determinant } (a_{ij}),$$

where the matrix $(a_{ij})$ is given by $a_{ij} = B(x_i, x_j)$.

In order to verify that this is an inner product, one again needs to check that the adjoint $Ad_R(\Lambda^n B)$ is an isomorphism, [B 30]. We then call $(\Lambda^n M, \Lambda^n B)$ the discriminant inner product space.

Comment 3.6 The adjoint map of B, $Ad_R B$ or simply Ad B, Ad B: $M \to Hom_D(M,I)$ is an isomorphism by hypothesis. Taking the $n^{\underline{th}}$ exterior power, $\Lambda^n(\text{Ad } B): \Lambda^n M \to \Lambda^n(Hom_D(M,I))$ is an isomorphism.

However, $Hom_D(M,I) \simeq \bigoplus_{i=1}^{n-1} Hom_D(D,I) \oplus Hom_D(A,I) \simeq \bigoplus_{i=1} I \oplus \bar{A}^{-1}I$.

Thus $\Lambda^n(Hom_D(M,I)) \simeq I \otimes \ldots I \otimes \bar{A}^{-1}I$. However, $I \otimes \ldots \otimes I \otimes \bar{A}^{-1}I$ is not in general isomorphic to $Hom_D(\Lambda^n M, I)$ so that $\Lambda^n B$ is not in general non-singular.

However, for I = D, the Dedekind ring of integers,

$$Ad(\Lambda^n B) = \Lambda^n(\text{Ad } B): \Lambda^n M \to Hom_D(\Lambda^n M, D)$$

will be an isomorphism by the above, and $(\Lambda^n M, \Lambda^n B)$ is indeed an inner product space.

We note that M is free as a D-module if and only if the ideal is principal. Thus the discriminant inner product space yields information about the structure of the D-module M.

We may apply the operation of 3.4, tensoring with the quotient field, to the discriminant inner product space. $\Lambda^n B \otimes 1$ is then multiplication by a fixed $x_0 \in E$. Thus associated with an inner product space (M,B) is a pair $(x_0, A)$, where $A = \Lambda^n M$, and $x_0$ is as described.

This $x_0$ specifies the adjoint isomorphism,

$$Ad_R \Lambda^n B: \quad A \to \mathrm{Hom}_D(A,D) = \bar{A}^{-1}D, \text{ by } \Lambda^n B(a_1, a_2) = x_0 a_1 \bar{a}_2 ,$$

with $x_0$ unique in $F^*/NE^*$, where $F^*$ is the fixed field F of $-$, and $NE^*$ denotes the multiplicative group of norms of elements in $E^*$. Hence, $x_0 A = \bar{A}^{-1}D$, ie. $x_0 A\bar{A} = D$.

## 4. The symmetry operator

In Definitions 2.1 and 2.2 of an inner product space, and non-singular mapping, we made use of the right adjoint operator $Ad_R$. In this section, we relate the two adjoint operators $Ad_R$ and $Ad_L$.

Using Theorems 2.6 and 2.9, this is done as follows.

**Theorem 4.1** *Let (M,B) be an inner product space of either type (a) or (b). Let $B: M \times M \to K$ satisfy $B(dx,y) = B(x,\bar{d}y) = dB(x,y)$. Then the right adjoint map $Ad_R B$ is an isomorphism if and only if the left adjoint map is.*

*Proof:* Let $\phi$ denote the canonical isomorphism

$$\phi : \quad M \to \mathrm{Hom}_D(\mathrm{Hom}_D(M,K),K)$$

of Theorems 2.6 and 2.9, given by $\phi(x)(f) = \overline{f(x)}$.

Assume $Ad_R B$ is an isomorphism. We can thus identify $M \simeq Hom_D(M,K)$ via $Ad_R B$. Here $y \in M$ is identified with $B(-,y) \in Hom_D(M,K)$.

The isomorphism $\phi$ is now given by:

$$\phi: \quad M \to Hom_D(Hom_D(M,K),K) \xrightarrow{Ad_R B} Hom_D(M,K)$$
$$x \to \phi_x \quad \text{where } \phi_x(y) = \overline{B(x,y)} \; ,$$

in other words $\phi$ is $Ad_L B$. The converse follows similarly. $\square$

Corollary 4.2 Let $(M,B)$ be an inner product space. Then we can define a unique D-linear isomorphism $s: \quad M \to M$ by the equation $\overline{B(x,y)} = B(y,sx)$.

Proof: $Ad_L B(x) = \overline{B(x,-)} \in Hom_D(M,K)$. Since B is an inner product, we define $s(x)$ by :

$$Ad_R(sx) = \overline{B(x,-)} = B(-,sx).$$

$s$ is an isomorphism by Theorem 4.1. $\square$

Notation 4.3 We shall reserve the letter s for this map which is related to the symmetry of B as described above. It is precisely this map s which enables us to work with non-symmetric inner product spaces.

Let N be a subspace of M. We say that N is s invariant provided

$s(N) \subset N$.

<u>Proposition 4.4</u> <u>N is</u> s <u>invariant if and only if</u> sN = N.

<u>Proof</u>: Sufficiency is clear.

In order to prove necessity, suppose $s(N) \subset N$. Then we can form an ascending chain of submodules of M, $T_i = \{m \in M: s^i(m) \in N\}$, $T_{i+1} \supseteq T_i$. Since D is Noetherian, and M is finitely generated, M is Noetherian, [S 47]. Hence, this chain terminates. Suppose $T_i = T_N$ for $i \geq N$.

Claim: $T_0 = T_1 = \ldots = T_N$, and hence sN = N.

It clearly suffices to show that $T_i \neq T_{i+1}$ implies $T_{i+1} \neq T_{i+2}$. Suppose then that $T_i \neq T_{i+1}$. Let $m_0 \in T_{i+1} - T_i$. Then $s^{i+1}(m_0) \in N$ and $s^i(m_0) \notin N$. s is an isomorphism, so there exists $m_1 \in M$ with $sm_1 = m_0$. Hence, $m_1 \in T_{i+2}$. If $m_1 \in T_{i+1}$, then $m_0 \in T_i$ which is a contradiction. Thus, $m_1 \in T_{i+2} - T_{i+1}$. □

We observe that Proposition 4.4 can be proved for any ring R in place of D, with R not necessarily Noetherian. Namely, let s be an isomorphism s: $M \to M$ of a finitely generated R-modules, with R not necessarily Noetherian.

Claim: If $N \subset M$ is a submodule with $sN \subset N$, then sN = N.

We sketch the proof using the following form of Nakayama's Lemma: Assume M is finitely generated. Then M = IM implies I + Ann(M)= R where Ann(M)denotes the annihilator of M.

Consider the diagram :

$$\begin{array}{ccccccc} 0 & \rightarrow & M & \xrightarrow{s} & M & \rightarrow & 0 \\ & & \downarrow & & \downarrow & & \\ & & M/N & \xrightarrow{\hat{s}} & M/N & \rightarrow & 0 \end{array}$$

It suffices to show $\hat{s}$ is 1-1. View $M/N$ as an $R[x]$-module, where $x$ acts as $\hat{s}$ . Suppose $x\, m_0 = 0$. We want to show $m_0 = 0$. Let $I = (x)$. Since $\hat{s}$ is onto, $M/N = I(M/N)$. Hence, by Nakayama, above, $I + \text{Ann}\,(M/N) = R[x]$. Write $1 = rx + t$, where $r \in R[x]$, $t \in \text{Ann}\,(M/N)$. Then $1 \cdot m_0 = rx \cdot m_0 + t \cdot m_0 = 0$, so $m_0 = 0$. Thus $\hat{s}$ is 1-1. □

For a subspace $N$ of $M$, we define

$$N_R = \{v \in M\colon B(n,v) = 0 \text{ for all } n \in N\} .$$

Thus $N_R$ is the kernel of $\text{Ad}_R B$ restricted to $N$, $\text{Ad}_R B\colon\ M \rightarrow \text{Hom}_D(N,K)$. We call $N_R$ the right orthogonal complement of $N$. The underlying inner product space is understood. Similarly,

$$N_L = \{v \in M\colon B(v,n) = 0 \text{ for all } n \in N\} .$$

Now, let $N$ be $s$ invariant. Since $sN = N$, it follows that $N_L = N_R$. We denote this common orthogonal complement $N^{\perp}$ .

<u>Note 4.5</u> $N^{\perp}$ <u>is only defined when</u> $N$ <u>is</u> $s$ <u>invariant</u>.

*Proposition 4.6* *If* (M,B) *is an inner product space of type* (a), *meaning* M *is torsion free, then* $N_L$ *and* $N_R$ *are direct summands of* M.

*Proof*: Consider the exact sequence:

$$0 \to N_L \to M \to M/N_L \to 0 .$$

It suffices to show that $M/N_L$ is torsion free, hence projective over the Dedekind domain D. Then the sequence splits and $N_L$ is a summand.

Suppose to the contrary $M/N_L$ has torsion. Then there exists $x \in M$, $d \neq 0$, $d \in D$, with $x \in N_L$ and $dx \in N_L$. So $B(dx,y) = dB(x,y) = 0$ for all $y \in N$. Thus $B(x,y) = 0$ for all $y \in N$ since D is a domain. Hence $x \in N_L$, contradiction. Thus $M/N_L$ is torsion free. □

We have already remarked that if N is s invariant, then $N_L = N_R$. The above shows that $N_L$ and $N_R$ are always summands. These two conditions turn out to give the converse.

*Proposition 4.7* *Let* (M,B) *be an inner product space with* M *torsion free. Let* N *be a summand of* M. *Then* N *is* s *invariant if and only if* $N_L = N_R$.

*Proof*: As observed before Note 4.5, necessity is clear. In order to prove sufficiency, consider the two exact sequences:

$$0 \to N_L \to M \xrightarrow{Ad_L B} \mathrm{Hom}_D(N,K) \to 0$$

$$0 \to (N_L)_R \to M \xrightarrow{Ad_R B} \mathrm{Hom}_D(N_L,K) \to 0 .$$

$Ad_L B$ and $Ad_R B$ are onto because N is a summand. Clearly we have rank $N_L$ = rank $(Hom_D(N_L,K))$. Thus, by the two sequences above, rank $(N_L)_R$ = rank $(Hom_D(N,K))$ = rank N. However, $B(n_\ell,n) = 0$ for all $n_\ell \in N_L$, $n \in N$. Thus $N \subset (N_L)_R$. Since N is a summand, ranks equal, it follows that $N = (N_L)_R$. Similarly, $N = (N_R)_L$.

We now wish to show that $sN \subset N = (N_L)_R$. So we compute $B(n_\ell,sn) = \overline{B(n,n_\ell)} = 0$ for all $n_\ell \in N_L$, since $N_L = N_R$ by hypothesis. Thus $sN \subset (N_L)_R = N$ as desired. □

<u>Proposition 4.8</u> $N = N_L$ <u>if and only if</u> $N = N_R$.

<u>Proof</u>: Suppose $N = N_L$. Then $B(m,n) = 0$ for all $m, n \in N$. Thus, $n \in N_R$ and $N \subset N_R$. By Proposition 4.6, $N = N_L$ implies N is a summand. Clearly, as in Proposition 4.7, rank N = rank $N_R$, so that $N = N_R$.

The converse is similar. □

Theorem 4.1 stated that for an inner product space (M,B), both $Ad_R B$ and $Ad_L B$ are isomorphisms. This enabled us to define the symmetry operator s, with $\overline{B(x,y)} = B(y,sx)$. In a like manner, one can see:

<u>Proposition 4.9</u> <u>Fixing an inner product space</u> (M,B), <u>let</u> $\ell$: $M \to M$, <u>be a</u> D-<u>linear operator</u>. <u>Then there is a unique</u> $\ell^*$: $M \to M$, D-<u>linear</u>, <u>with</u> $B(x,\ell y) = B(\ell^* x,y)$.

Notation: $\ell^*$ is called the adjoint operator of $\ell$, not to be confused with the adjoint maps $Ad_R B$, $Ad_L B$ previously defined.

Proof: For fixed x, we have the map $\overline{B(x,\ell(-))} \in \text{Hom}_D(M,K)$. Since B is non-singular, $\text{Ad}_L$ is an isomorphism and we can find a unique $w \in M$ such that:

$$\overline{B(x,\ell(-))} = \overline{B(w,-)} .$$

Define $\ell^*x = w$. Then $B(x,\ell y) = B(\ell^*x,y)$. $\ell^*$ is clearly a well-defined, D-linear map with the desired properties. $\text{Ad}_L$ being an isomorphism shows $\ell^*$ is unique. □

One can similarly define a left adjoint operator of $\ell$, $^*\ell$ by insisting $B(\ell x,y) = B(x,{}^*\ell y)$. It then follows that if $\ell \in \text{Hom}_D(M,M)$

$$B(({}^*\ell)^*x,y) = B(x,{}^*\ell y) = B(\ell x,y).$$

Since B is non-singular, $({}^*\ell)^* = \ell$. Similarly, ${}^*(\ell^*) = \ell$, so that these two * operations are inverses of one another.

Further, we note that

$$\begin{aligned} B(\ell^{**}x,y) &= B(x,\ell^*y) \\ &= B(\ell^*y,sx) \\ &= B(y,\ell sx) \\ &= B(s^{-1}\ell sx,y) \end{aligned}$$

Thus $\ell^{**} = s^{-1}\ell s$. Also we have

$$B(x,y) = B(y,sx) = B(sx,sy) = B(s^*sx,y)$$

so that $s^* = s^{-1}$. Similarly $^*s = s^{-1}$. We summarize these remarks in

Theorem 4.10 *The correspondences* $\ell \to \ell^*$ *and* $\ell \to {}^*\ell$ *of the algebra of linear operators on M are inverses. They satisfy* $\ell^{**} = s^{-1}\ell s$, $s^* = {}^*s = s^{-1}$.

Thus $\ell^{**} = \ell$ for all $\ell \in Hom_D(M,M)$ if and only if s is central.

An easy calculation also shows $(\ell\tau)^* = \tau^*\ell^*$, so that when s is central the correspondence $\ell \to \ell^*$ gives an anti-involution of the algebra of linear operators on M.

## 5. The Witt equivalence relation

Definition 5.1 *Let* k *be given,* $k \in D$. *A degree* k *mapping structure over* D *is a triple* $(M,B,\ell)$ *satisfying:*

(a) $(M,B)$ *is an inner product space over* D .

(b) $\ell: M \to M$ *is a D-linear map satisfying*

$B(\ell x,\ell y) = kB(x,y)$ *for all* $x,y \in M$ .

$\ell$ is called a *map of degree* k. For all future considerations, we shall assume henceforth that $k \in Z$.

In the case that M is torsion free, and $k \neq 0$, it follows that $\ell$ is non-singular. To see this suppose $\ell(x) = 0$. Then $B(\ell x, \ell y) = kB(x,y) = 0$. Since $B$ has values in $K = I$ a - invariant fractional ideal, we cancel $k$ and conclude $B(x,y) = 0$ for all $y \in M$. However, $B$ is non-singular, so that $x = 0$, and $\ell$ is 1-1.

The Witt equivalence relation for degree k mapping structures comes from:

*Definition 5.2 A degree k mapping structure* $(M,B,\ell)$ *is metabolic if there is a D-submodule* $N \subset M$ *satisfying*:

(a) $N$ *is* $\ell$ *invariant*

(b) $N$ *is* $s$ *invariant*

(c) $N = N^{\perp}$

When $(M,B,\ell)$ is metabolic, an $N$ satisfying (a), (b) and (c) above will be called a *metabolizer* for $M$. We shall also refer to the triple $(M,B,\ell)$ as $M$, when $B$ and $\ell$ are understood, and speak of $M$ as being metabolic.

The operation of direct sum on inner product spaces extends to degree $k$ mapping structures. The notation:

$$(V,B,\ell) \oplus (W,B',\ell') = (V \oplus W, B \oplus B', \ell \oplus \ell').$$

It is clear that $\ell \oplus \ell'$ is of degree $k$ with respect to $B \oplus B'$

At this point we can introduce a relation $\sim$ on degree $k$ mapping structures by:

$$(V,B,\ell) \sim (W,B',\ell') \quad \text{when} \quad (V \oplus W, B \oplus -B', \ell \oplus \ell')$$

is metabolic. In what follows, we will show $\sim$ is an equivalence relation, called the Witt equivalence relation. This agrees with the usual notion of Witt equivalence [M-H] when no $\ell$ is present.

Notation: $W^{+1}(k,K)$ denotes degree $k$ mapping structures $(M,B,\ell)$ modulo $\sim$ with values in $K$; together with the additional requirement that the symmetry operator $s$ is the identity map, so $B$ is symmetric. The underlying ring $D$ is understood. Emphasis is given to the range of the inner product, namely $K$.

Similarly, $W^{-1}(k,K)$ is Witt equivalence classes of triples $(M,B,\ell)$ having $B$ skew-symmetric.

When there is no $\ell$, so that we are taking inner product spaces modulo $\sim$, without condition (a) of 5.2, we denote the symmetric equivalence classes $W^{+1}(K)$, the skew-symmetric $W^{-1}(K)$. Finally, when no $\ell$ is present, with no symmetry requirement at all placed on $B$, the resulting Witt group is denoted $A(K)$.

We have also defined the notion of $B$ being u Hermitian. We write $H_u(K)$ to denote those Witt equivalence classes $[M,B]$ for which $B(x,y) = u\overline{B(y,x)}$.

In summary, our notation is:

$W^{+1}$ : $B$ symmetric

$W^{-1}$ : $B$ skew-symmetric

$A$ : no symmetry requirements on $B$ ($B$ asymmetric)

$H_u$ : $B$ u Hermitian

If we write $W^{+1}(K)$, we are thinking of pairs $(M,B)$; if we write $W^{+1}(k,K)$, we are thinking of triples $(M,B,\ell)$ with $B(\ell x,\ell y) = kB(x,y)$.

The $K$ means that $B: M \times M \to K$.

We write $A(k,K)$ to denote the group which consists of triples $(M,B,\ell)$, with no symmetry requirements on $B$.

If we let $k$ range over $\mathbb{Z}$, we can form a graded ring, with multiplication defined by $\otimes$,

$$A(k,K) \times A(k',K') \to A(kk',KK')$$
$$(M,B,\ell) \times (M',B',\ell') \to (M \otimes_D M', B \otimes B', \ell \otimes \ell')$$

This follows from 3.2.

However, for this paper, we shall only be concerned with the Abelian group structure arising from direct sum.

Our objective now is to show that $\sim$ is an equivalence relation. It is easy to see that $\sim$ is reflexive, since $(V \oplus V, B \oplus -B, \ell \oplus \ell)$ has metabolizer $N = \{(x,x): x \in V\}$. Clearly $\sim$ is symmetric. We must show $\sim$ is transitive.

Again when $(M,B,\ell)$ is metabolic, we will say $M$ is metabolic, the $B$, $\ell$ being understood, and write $M \sim 0$. The following proposition is clear.

<u>Proposition 5.3</u> $\sim$ <u>is transitive if and only if</u> $H \sim 0$ <u>and</u> $M \oplus H \sim 0$ <u>implies</u> $M \sim 0$.

We call $M$ stably metabolic if there exists $H \sim 0$ with $M \oplus H \sim 0$. We may then restate Proposition 5.3 as saying $\sim$ is transitive if and only if stably metabolic implies metabolic.

<u>Comment</u>: Once we have show that $\sim$ is transitive, it is also clear that the following relation, $\approx$, would have yielded the same

relation as $\sim$ . Define $M_0 \approx M_1$ if and only if there exists $H_0 \sim 0$, $H_1 \sim 0$ with $M_0 \oplus H_0$ isomorphic to $M_1 \oplus H_1$ .

__Lemma 5.4__ __Suppose__ $M$ __is a finitely generated torsion free D-module. Then__ $(M,B,\ell) \sim 0$ __over__ $D$ __if and only if__ $(M,B,\ell) \otimes_D E \sim 0$ __over__ $E$ .

Here $E$ is the quotient field of $D$ (see 3.4).

__Proof__: Necessity is clear, for if $N$ is a metabolizer for $(M,B,\ell)$, then $N \otimes_D E$ is a metabolizer for $(M,B,\ell) \otimes_D E$.

To show sufficiency, consider the exact sequence

$$0 \to D \to E \to E/D \to 0 .$$

Tensoring with $M$, we note that $M$ is embedded into $M \otimes_D E$ as $M \otimes 1$. Suppose $M \otimes_D E$ has metabolizer $N = N^{\perp}$. Let

$$N_1 = N \cap (M \otimes 1) \subset M \otimes 1 .$$

Claim: $N_1 = N_1^{\perp}$ in $M \otimes 1 \cong M$, so that $M \sim 0$.

To begin with, $N_1$ is $s,\ell$ invariant since $N$ is. It also is clear that $N_1 \subset N_1^{\perp}$. Conversely, if $x \otimes 1 \in N_1^{\perp}$, $B(x \otimes 1, y \otimes 1) = 0$ for all $y \otimes 1 \in N_1$. However, if $y \otimes r \in N$, then $y \otimes 1 \in N_1$, so $B(x \otimes 1, y \otimes r) = \bar{r}B(x \otimes 1, y \otimes 1) = 0$. Hence $x \otimes 1 \in N^{\perp} = N$. Thus $x \otimes 1 \in N_1$, and $N_1^{\perp} \subset N_1$. Thus $N_1 = N_1^{\perp}$ and $M \otimes 1 \cong M \sim 0$ . □

Using Lemma 5.4, we see that to prove $\sim$ is transitive in the case that $M$ is torsion free, we can assume that $D = E$ a field, and $K = E$ .

We thus assume for the rest of the proof of transitivity that $K = E = D$ a field when $M$ is torsion free, and $K = E/I$ for $M$ torsion as usual.

In either case $K$ is an injective D-module and we have:

<u>Theorem 5.5</u> <u>Let</u> $(M,B)$ <u>be an inner product space, with values in</u> $K = E$ <u>or</u> $K = E/I$. <u>If</u> $N$ <u>is</u> $\ell$, s <u>invariant, then</u> $N = (N^{\perp})^{\perp}$ .

<u>Remark 5.6</u> <u>This is not true for</u> $K = I$ <u>an arbitrary fractional ideal or even</u> $K = D$ .

<u>Proof</u>: We have the exact sequence

$$0 \to N^{\perp} \to M \xrightarrow{Ad_R B} \mathrm{Hom}_D(N,K) \to 0$$

$Ad_R B$ is onto since $K$ is an injective D-module.

Applying the Hom functor, we obtain

$$0 \to \mathrm{Hom}_D(\mathrm{Hom}_D(N,K),K) \to \mathrm{Hom}_D(M,K) \to \mathrm{Hom}_D(N^{\perp},K) \to 0$$

Again, $\mathrm{Ext}(\mathrm{Hom}_D(N,K),K) = 0$ , since $K$ is injective and the last map is onto.

We can identify $Hom_D(Hom_D(N,K),K) \simeq N$ by Theorem 2.6. This clearly yields the commutative diagram:

$$0 \rightarrow N \rightarrow Hom_D(M,K) \rightarrow Hom_D(N^{\perp},K) \rightarrow 0$$
$$\downarrow \qquad\qquad \downarrow \qquad\qquad\qquad \downarrow$$
$$0 \rightarrow (N^{\perp})^{\perp} \rightarrow M \xrightarrow{Ad_R B} Hom_D(N^{\perp},K) \rightarrow 0$$

The inner product provides an isomorphism of the middle terms, so by diagram chase [M 50] the inclusion $N \subset (N^{\perp})^{\perp}$ is an isomorphism.

**Lemma 5.7** For any two $\ell$, s invariant submodules R and S of M, where (M,B) is an inner product space as above, we have:

(1)  $(R + S)^{\perp} = R^{\perp} \cap S^{\perp}$

and

(2)  $R^{\perp} + S^{\perp} = (R \cap S)^{\perp}$

**Proof:** (1) follows from the definition of $\perp$. To show (2), observe that

$$(R^{\perp} + S^{\perp})^{\perp} = (R^{\perp})^{\perp} \cap (S^{\perp})^{\perp} = R \cap S .$$

Thus, taking $\perp$, $R^{\perp} + S^{\perp} = (R \cap S)^{\perp}$.  $\square$

**Lemma 5.8** Let (M,B) be an inner product space as above. Suppose that $M \sim 0$ with metabolizer N. Let $L \subset M$ satisfy $L \subset L^{\perp}$. Then $L + N \cap L^{\perp} = (L + N \cap L^{\perp})^{\perp}$.

**Remark 5.9** This Lemma shows how to go from a metabolizer N, and a subspace $L \subset L^{\perp}$ to another metabolizer, namely $L + N \cap L^{\perp}$, which contains the self annihilating subspace L.

Proof: The assumption that L is $\ell$, s invariant is understood, in order that $L^{\perp}$ make sense.

We compute using Lemma 5.7:

$$\begin{aligned}(L + (N \cap L^{\perp}))^{\perp} &= L^{\perp} \cap (N \cap L^{\perp})^{\perp} \\ &= L^{\perp} \cap (N^{\perp} + (L^{\perp})^{\perp}) \\ &= L^{\perp} \cap (N + L) \quad \text{since} \quad N^{\perp} = N \quad \text{and} \quad (L^{\perp})^{\perp} = L \\ &= (L^{\perp} \cap N) + (L^{\perp} \cap L) \\ &= L + (N \cap L^{\perp}) \quad \text{since} \quad L \subset L^{\perp}.\end{aligned}$$

Thus $L + (N \cap L^{\perp})$ is also a metabolizer. □

Theorem 5.10 (Transitivity of ~) Let $H \sim 0$. Then $M \oplus H \sim 0$ if and only if $M \sim 0$.

Proof: Sufficiency is clear. To show necessity, let N be a metabolizer for $M \oplus H$, and $H_0$ a metabolizer for H. We embed $H_0$ and H into $M \oplus H$ as $0 \oplus H_0$, $0 \oplus H$ respectively. Notice that $0 \oplus H_0 \subset (0 \oplus H_0)^{\perp}$, so that by Lemma 5.8, we may rechoose N such that $0 \oplus H_0 \subset N$.

We review our notation. (M,B), (H,B') and $(M \oplus H, B \oplus B')$ are the inner product spaces. We will write elements in $M \oplus H$ as pairs (x,y) with $x \in M$, $y \in H$.

Let $N_0$ = projection of N onto $M = \{a \in M: (a,h) \in N \text{ for some } h\}$.

Claim: $N_0$ is a metabolizer for $M$.

$N_0$ is clearly $\ell$, $s$ invariant since $N$ is, and projection commutes with $\ell$, $s$ on $M \oplus H$.

We first show that $N_0 \oplus H_0 = N$. If $(a,h) \in N$, we claim $h \in H_0$. Let $(0,h_1) \in 0 \oplus H_0 \quad N = N^{\perp}$. Then

$$\begin{aligned}(B \oplus B')((0,h),(0,h_1)) &= B(0,0) + B'(h,h_1) \\ &= B(a,0) + B'(h,h_1) \\ &= (B \oplus B')((a,h),(0,h_1)) \\ &= 0\end{aligned}$$

since $N = N^{\perp}$. Hence $B'(h,h_1) = 0$ for all $h_1 \in H_0$, so that $h \in H_0^{\perp} = H_0$ as claimed. Thus $N_0 \oplus H_0 = N$.

Clearly, $N_0 \subset N_0^{\perp}$. Conversely, let $b \in N_0^{\perp}$. Then by computing as above $(b,0) \in N^{\perp} = N$, so that $b \in N_0$. Hence $N_0 = N_0^{\perp}$ is a metabolizer for $M$. $\square$

Thus $\sim$ is an equivalence relation, and we may form the Witt group consisting of equivalence classes of triples $(M,B,\ell)$ modulo $\sim$. $(M,B,\ell) \sim (M_1,B_1,\ell_1)$ provided $(M \oplus M, B \oplus -B_1, \ell \oplus \ell_1) \sim 0$.

Notation: $[M,B,\ell]$ will denote the Witt equivalence class of $(M,B,\ell)$.

## 6. Anisotropic representatives

Our final goal of this chapter is to find a representative of each equivalence class. As long as $K = E$ a field, or $E/I$ in the torsion case, this representative is unique.

We begin by describing the representative we will obtain.

Definition 6.1 A degree $k$ mapping structure $(M,B,\ell)$ is anisotropic if for any $s,\ell$ invariant D-submodule $N$ of $M$, $N \cap N^{\perp} = 0$.

Theorem 6.2 Every Witt equivalence class $[M,B,\ell]$ has an anisotropic representative.

We prove this theorem by way of a sequence of Lemmas which are of interest in their own right.

Lemma 6.3 Let $T$ be an $s,\ell$ invariant D-submodule of $M$, with $T \subset T^{\perp}$. Then $T^{\perp}/T$ inherits a quotient degree $k$ mapping structure $(T^{\perp}/T,\bar{B},\bar{\ell})$.

Proof: Let $[t]$ denote an element in $T^{\perp}/T$. Define

$$\bar{B}([t_1],[t_2]) = B(t_1,t_2)$$

where $t_1$, $t_2$ are representatives of $[t_1]$, $[t_2]$ respectively. $\bar{B}$ is clearly well-defined since $T$ is self-annihilating, ie. $T \subset T^{\perp}$. It is likewise clear that $\bar{\ell}$, the induced map on $T^{\perp}/T$ is of degree $k$ with respect to $\bar{B}$, and well-defined.

We must show that $\bar{B}$ is an inner product, ie. that

$\mathrm{Ad}_R\bar{B}$: $T^{\perp}/T \to \mathrm{Hom}_D(T^{\perp}/T,K)$ is an isomorphism.

Applying the functor $\mathrm{Hom}_D(-,K)$ to the exact sequence:

$$0 \to T \to T^{\perp} \to T^{\perp}/T \to 0,$$

we obtain the embedding;

$$0 \to \mathrm{Hom}_D(T^{\perp}/T,K) \to \mathrm{Hom}_D(T^{\perp},K).$$

Suppose $\bar{g} \in \mathrm{Hom}_D(T^{\perp}/T,K) \to \tilde{g} \in \mathrm{Hom}_D(T^{\perp},K)$. We can lift $\tilde{g}$ to $g\colon M \to K$ since $T^{\perp}$ is a summand by Proposition 4.6 in the torsion free case and since $K$ is injective in the torsion case.

$\mathrm{Ad}_R B\colon M \to \mathrm{Hom}_D(M,K)$ is an isomorphism. Hence $g = B(-,x)$. $g$ restricted to $T$ equals 0, so $x \in T^{\perp}$. Thus $(x + t)$ gives the same $\bar{g}$ for all $t \in T$. So we may read $x \in T^{\perp}/T$.

This procedure defines a map:

$$\mathrm{Hom}_D(T^{\perp}/T,K) \to T^{\perp}/T, \text{ namely } \bar{g} \to \tilde{g} \to [x].$$

The inverse of this map is simply

$$[x] \in T^{\perp}/T \xrightarrow{\mathrm{Ad}_R\bar{B}} \bar{B}(-,[x]).$$

Hence $\mathrm{Ad}_R\bar{B}$ is an isomorphism and $\bar{B}$ is an inner product. □

<u>Lemma 6.4</u> <u>With the same hypotheses as in Lemma</u> 6.3, $M \oplus -T^{\perp}/T$ <u>is metabolic</u>.

<u>Proof</u>: In the torsion free case, by Lemma 5.4, we may assume

$K = E$ a field. Thus, in any case, there is no loss of generality in assuming the hypotheses of Theorem 5.5, namely that $K = E$ or $K = E/I$. Consequently, for $N$ an $\ell$, s invariant subspace of $M$, we have $N = (N^{\perp})^{\perp}$.

We wish to show $M \oplus -T^{\perp}/T \sim 0$. So consider $N = \{(x, x + T) : x \in T^{\perp}\}$. $N$ is an $\ell$, s invariant subspace, and clearly $N \subset N^{\perp}$. Let $(a, b + T) \in N^{\perp}$, with $b \in T^{\perp}$. We compute

$$(B \oplus -\bar{B})\big((a, b + T), (x, x + T)\big) = B(a,x) - B(b,x) = 0$$

for all $(x,x+T) \in N$. Thus $B(a - b,x) = 0$ for all $x \in T$. Hence, $(a - b) \in (T^{\perp})^{\perp} = T$, since by assumption Theorem 5.5 applies. So $[b] = [a] - [(a - b)] = [a]$ in $T^{\perp}/T$, and $(a,b + T) = (a,a + T) \in N$. Therefore, $N^{\perp} \subset N$, and $N$ is a metabolizer for $M \oplus -T/T$.

Lemma 6.4 shows then that $M \sim T^{\perp}/T$ whenever $T \subset T^{\perp}$. In the torsion free case, we can use that rank $M < \infty$ to conclude, after successive applications of Lemmas 6.3 and 6.4, that $M \sim M_0$ where $M_0$ has no $\ell$,s invariant subspace $T$ with $T \subset T^{\perp}$. In other words, $M_0$ is anisotropic.

For the torsion module case, we repeatedly apply Lemmas 6.3 and 6.4 to obtain sequences:

$$M \supset T^{\perp} \supset T_1^{\perp} \cdots \quad \cdots \supset T_2 \supset T_1 \supset T$$

Since $M$ is Noetherian, the ascending chain condition implies that the sequence $\{T_i\}$ terminates. Hence, it follows the sequence $\{T_i\}$

will also terminate.

Since both chains terminate, we have $M \sim T_r^{\perp}/T_r$ ; with $T_r^{\perp}/T_r$ having no $\ell$, s invariant submodule $T_{r+1}$ with $T_{r+1} \subset T_{r+1}^{\perp}$. Thus, $[M,B,\ell]$ has an anisotropic representative, $T_r^{\perp}/T_r$, as claimed. This completes the proof of Theorem 6.2.

This anisotropic representative need not be unique for torsion free D-modules [M-H 28]. However, for $K = E$ a field, or $K = E/I$, we shall see that it is unique.

<u>Theorem 6.5</u> <u>As long as</u> $[M,B,\ell] \in W(k,K)$ <u>satisfies</u> <u>Theorem</u> 5.5, ie. <u>for</u> $K = E$ <u>or</u> $E/I$, <u>every Witt equivalence class</u> $[M,B,\ell]$ <u>has a unique anisotropic representative up to isomorphism</u>.

<u>Proof</u>: Suppose $(M,B,\ell) \sim (M',B',\ell')$, with $M$ and $M'$ both anisotropic. Let $N \subseteq M \oplus M'$ be a metabolizer, with respect to $B \oplus -B'$. We will show that $N$ is the graph of an isomorphism $f: M \to M'$ which satisfies $B'(f(k),f(y)) = B(x,y)$, $\ell' \circ f = f \circ \ell$ and $s' \circ f = f \circ s$. Thus $f$ is an isomorphism between $(M,B,\ell)$ and $(M',B',\ell')$.

Let $A = \{a \in M$: there exists $a_1 \in M'$ with $(a,a_1) \in N\}$.

Claim: Given $a \in A$, then there exists a unique $a_1 \in M'$ with $(a,a_1) \in N$.

For suppose $(a,a_1)$ and $(a,a_2) \in N$. Then $(0,a_1 - a_2) \in N$. Consider the $\ell'$, s' invariant subspace, $M_1$ of $M'$ generated by $a_1 - a_2$. Since $N$ is $s \oplus s'$ and $\ell \oplus \ell'$ invariant, this subspace

$M_1$ will have $(0,M_1) \subset N$. Hence $M_1 \subset M_1^{\perp}$ since $N = N^{\perp}$. This is a contradiction to $M'$ being anisotropic unless $a_1 - a_2 = 0$ so that $a_1 = a_2$.

Similarly, let $B = \{a_1 \in M'$: there exists $a \in M$ with $(a,a_1) \in N\}$. As above, each $a_1 \in B$ has a unique $a \in M$ with $(a,a_1) \in N$. It follows that $N$ is the graph of a 1-1 function $f$.

We claim that $A = M$ and $B = M'$. To see this we show that $A^{\perp} = 0$ in $M$, hence $(A^{\perp})^{\perp} = A = M$. So let $a \in A$, and consider $(a,0) \in M \oplus M'$. Take any $(x,y) \in N$. Then:

$$(B \oplus -B')((a,0),(x,y)) = B(a,x) = 0$$

since

$$a \in A^{\perp},\ x \in A.$$

Thus, $(a,0) \in N^{\perp} = N$. By the first claim, this implies $a = 0$. A similar argument shows $B = M'$. It follows that $f\colon M \to M'$ is an isomorphism.

Let $(a,f(a)) \in N$. Then $(\ell a, \ell' f(a)) \in N$ since $N$ is $\ell \oplus \ell'$ invariant. Thus, by definition, $(f \circ \ell)(a) = (\ell' \circ f)(a)$. Similarly, $(f \circ s)(a) = (s' \circ f)(a)$.

Finally, consider $(x,f(x))$ and $(y,f(y)) \in N$.

$$(B \oplus -B')((x,f(x)),(y,f(y)) = 0$$

so

$$B(x,y) - B'(f(x),f(y)) = 0 \quad \text{and} \quad B(x,y) = B'(f(x),f(y))$$

as desired. □

## Chapter II WITT INVARIANTS

Having defined the Witt ring we are led to examine invariants which will enable us in many cases to compute at least the group structure of $W(k,F)$.

Section 1 begins with a preliminary discussion of prime ideals and some results from algebraic number theory. In Section 2 we state the basic properties of Hilbert symbols. The reader should also see O'Meara [O'M] Introduction to Quadratic Forms for a complete exposition.

Following this introduction, we continue by considering $(M,B)$ a u Hermitian inner product space over a field $E$ with involution $-$ and fixed field $F$. $B: M \times M \to E$ satisfies:

$$B(x,dy) = u\overline{B(dy,x)} = \bar{d}B(x,y) \quad \text{for} \quad d \in E, \quad x,y \in M .$$

In Section 3, we discuss the rank mod 2 of $M$ as a Witt group invariant. Next, in Section 4, we introduce the discriminant invariant, the Witt analog of the determinanat for matrices. Thus, we review the matrix representation of $B$ and diagonalization in order to define this invariant. Section 5 is concerned with the signature invariants, which arise from the real infinite ramified primes.

These invariants completely determine the Hermitian group $H^{+1}(E)$ for $E$ an algebraic number field by Landherr's Theorem.

Notation: $F_2 = \{0,1\}$ = additive group of $Z$ modulo 2. (field with two elements)

$Z^* = \{1,-1\}$ = multiplicative group of units in $Z$.

## 1. Prime ideals

The setting is as in Chapter 1. Again $E$ is an algebraic number field together with an involution $-$. The fixed field of $-$ is $F$. The Dedekind rings of integers in $E$ and $F$ are denoted by $O(E)$ and $O(F)$ respectively. $O(F) = O(E) \cap F$.

If $\mathcal{P}$ is a prime ideal in $O(E)$, then $P = \mathcal{P} \cap O(F)$ will denote the corresponding prime ideal in $O(F)$.

Conversely, if $P$ is a prime ideal in $O(F)$, by going up, [A,Mc 63] there exists a prime ideal $\mathcal{P}$ in $O(E)$ with $\mathcal{P} \cap O(E) = P$. In fact, there may be several such prime ideals in $O(E)$ lying over $P$. The answer is given by considering $PO(E)$, [S 71]. We factor $PO(E) = \prod_{i=1}^{g} \mathcal{P}_i^{e_i}$. The $\mathcal{P}_i$ satisfy $\mathcal{P}_i \cap O(E) = P$. Since the extension $[E:F]$ is of degree 2, $\sum_{i=1}^{g} e_i f_i = 2$, where $f_i = [O(E)/\mathcal{P}_i : O(F)/P]$ is the residue field degree. We thus obtain the following cases:

1.1 Split $e = 1$, $f = 1$, $g = 2$. In this case $PO(E) = \mathcal{P}\bar{\mathcal{P}}$ where $\mathcal{P}$ is a prime ideal in $O(E)$ with $\mathcal{P} \neq \bar{\mathcal{P}}$. We say that $P$ splits in this case.

We may examine the split case in terms of the local completion of $F$ at $P$, which we denote $\tilde{F}(P)$.

Write $E = F(\sqrt{\sigma})$, and suppose $\sqrt{\sigma}$ satisfies the irreducible polynomial $p_{\sqrt{\sigma}/F}(x) = p(x)$. Then factor $p(x)$ in $\tilde{F}(P)[x]$. The split case corresponds to $p(x) = f_1(x) \cdot f_2(x)$. The prime spots $\mathcal{P}_i$ dividing $P$ are determined by $F$ - monomorphisms $\gamma\colon E \to L$, where $L$ is an algebraic closure of $\tilde{F}(P)$. The $-$ involution interchanges the prime spots, hence, $PO(E) = \mathcal{P}\bar{\mathcal{P}}$. $[\tilde{E}(\mathcal{P}):\tilde{F}(P)] = 1$ is the local degree.

1.2 Inert $e = 1$, $f = 2$. In this case $PO(E) = \mathcal{P}$ a prime in $O(E)$. $\mathcal{P} = \bar{\mathcal{P}}$, and we say $P$ remains prime, or is inert.

1.3 Ramified $e = 2$, $f = 1$. $PO(E) = \mathcal{P}^2$, $\mathcal{P} = \bar{\mathcal{P}}$ in this case also. We say $P$ ramifies.

In both the inert and ramified cases, $p(x) = p_{\sqrt{\sigma}/F}(x)$ is irreducible in $\tilde{F}(P)[x]$, and the local degree $[\tilde{E}(\mathcal{P}):\tilde{F}(P)]$ equals 2.

This describes the situation for finite primes.

We next consider all embeddings $\tau\colon F \to C$, where $C$ is the complex numbers. If $\tau\colon F \to R$ we call $\tau$ a real infinite prime. Otherwise $\tau$ is called a complex infinite prime. We denote infinite primes by $P_\infty$. Our only concern will be with real infinite primes.

Again, since $[E:F] = 2$, and the characteristic of these fields is 0 (not 2), we may write $E = F(\sqrt{\sigma})$, for $\sigma \in F^*$ unique up to multiplication by a square in $F^*$. For an infinite prime $P_\infty$, there are two cases:

1.4 Split If $P_\infty$ is complex infinite, $P_\infty$ is split. If $P_\infty$ is real infinite, and $\sigma > 0$ with respect to the ordering induced by $P_\infty$, we again say $P_\infty$ is split. In the case of a real split prime, $P_\infty$, the ordering of $F$ can be extended to $E$ in two distinct ways.

1.5 Ramified If $P_\infty$ is a real infinite prime, and $\sigma < 0$ with respect to the $P_\infty$ induced ordering we say $P_\infty$ is ramified. In this case, the $P_\infty$ ordering of $F$ can be uniquely extended to an embedding of $E$ into $C$ in such a way that the imaginary part of $\sqrt{\sigma}$ is positive.

Let $\tau$ denote the extension of $P_\infty$ to E. Then $\tau$ is equivariant with respect to complex conjugation $-$. This means there is a commutative diagram:

$$\begin{array}{ccc} E & \xrightarrow{\tau} & C \\ \downarrow - & & \downarrow - \\ E & \xrightarrow{\tau} & C \end{array}$$

The map $-: E \to E$ is the involution, $-: C \to C$ is complex conjugation. There should be no confusion.

Associated to a finite prime P in O(F), or $\mathcal{P}$ in O(E), is a discrete, non-Archimedian valuation $|\ |_P$, respectively $|\ |_{\mathcal{P}}$. $\mathcal{P}$ lies over P if and only if $|\ |_{\mathcal{P}}$ extends $|\ |_P$, as discussed in Chapter 1.

We next describe prime ideals in terms of local uniformizers. If $PO(E) = \mathcal{P}\bar{\mathcal{P}}$ is split, then a local uniformizer $\pi_P \in O_F(P)$ is also a local uniformizer for both $O_E(\mathcal{P})$ and $O_E(\bar{\mathcal{P}})$. Careful, this does not mean $\mathcal{P}$ and $\bar{\mathcal{P}}$ induce the same valuation. On the contrary, $O_E(\mathcal{P}) \neq O_E(\bar{\mathcal{P}})$. It only says $(\pi)$ in $O_E(\mathcal{P})$ is the unique maximal ideal.

If $PO(E) = \mathcal{P}$ is inert, then a local uniformizer $\pi_P$ for $O_F(P)$ is a local uniformizer for $O_E(\mathcal{P})$ also.

If $PO(E) = \mathcal{P}^2$ is ramified, then any local uniformizer $\pi$ of $O_E(\mathcal{P})$ will have norm $\pi\bar{\pi}$, and $\pi\bar{\pi}$ is a local uniformizer for $O_F(P)$. This follows since $\mathcal{P} = \bar{\mathcal{P}}$, hence $\pi$ and $\bar{\pi}$ are both local uniformizers for $O_E(\mathcal{P})$. So $v_{\mathcal{P}}(\pi\bar{\pi}) = 2 = v_{\mathcal{P}}(\pi_P)$. Thus $\pi\bar{\pi}$ is a local uniformizer for $O_F(P)$, when P ramifies.

For $y \in F^*$ , we summarize:

(1) If $P$ splits, $PO(E) = \mathcal{P}\bar{\mathcal{P}}$. $v_P(y) = v_{\mathcal{P}}(y)$
$= v_{\bar{\mathcal{P}}}(y)$

(2) If $P$ is inert, $PO(E) = \mathcal{P}$ $v_P(y) = v_{\mathcal{P}}(y)$

(3) If $P$ ramifies, $PO(E) = \mathcal{P}^2$ $2v_P(y) = v_{\mathcal{P}}(y)$

This is _not_ true for $y \in E^*$.

Associated to a prime $\mathcal{P}$ , finite or infinite, lying over a prime $P$, is the extension of localized completions. The degree $[\tilde{E}(\mathcal{P}):\tilde{F}(\mathcal{P})]$ is denoted $n_{\mathcal{P}}$ . $n_{\mathcal{P}} = 1$ if $\mathcal{P}$ is split and 2 otherwise.

## 2. Hilbert symbols

We begin by recalling the theorem of Hasse.

_Theorem 2.1_ _Let_ $y \in F^*$. _Then_ $y$ _is a norm from_ $E^*$ _if and only if_ $y \in \tilde{F}(\mathcal{P})^*$ _is a norm from_ $E(\mathcal{P})^*$, _for all_ $\mathcal{P}$ , _finite and infinite in_ $E$ [O'M 186].

This condition is trivial over $P$ split, for then $\tilde{F}(P) = \tilde{E}(\mathcal{P})$.

We should like to rephrase this in terms of Hilbert symbols [O'M 169]. We now state briefly the salient properties of these symbols.

If $a, \sigma \in F^*$, a symbol $(a,\sigma)_P$ is defined by: $(a,\sigma)_P = +1$ if and only if $a$ is a norm from $\tilde{F}(P)(\sqrt{\sigma})$ if and only if there exists $x, y \in \tilde{F}(P)$ satisfying $ax^2 + \sigma y^2 = +1$.

In terms of the prime ideals, we summarize.

**2.2** If P splits: $(a,\sigma)_P = +1$

**2.3** If P is inert: The local degree $n_P = [\tilde{E}(P):\tilde{F}(P)] = 2$.

By [O'M 169], evert local unit is a local norm, and the local uniformizer $\pi \in \tilde{F}(P)$ is not a local norm.

In terms of Hilbert symbols, for $a \in F^*$, we have the following. Let $a = \pi^n v$, for $\pi$ a local uniformizer and v a local unit.

$$\begin{aligned}(a,\sigma)_P &= (\pi,\sigma)_P^n (v,\sigma)_P \\ &= (\pi,\sigma)_P^n \\ &= (-1)^n \\ &= (-1)^{v_P(a)}\end{aligned}$$

**2.4** If P is ramified: Again the local degree is 2. As we have seen, in this case we may pick a local uniformizer $\pi_P$ of P to be a local norm, namely, $\pi_P = \pi_{\mathcal{P}} \bar{\pi}_{\mathcal{P}}$, where $\pi_{\mathcal{P}}$ is a uniformizer for $E(\mathcal{P})$. We thus study the local units.

The residue field, $O_F(P)/m(P) \simeq O(F)/P$ is isomorphic to the completion $\tilde{O}_F(P)/\tilde{m}(P)$. If u is a local unit, then for the following we denote by $u_1$ the image of u in the residue field, $O_F(P)/m(P)$.

<u>Claim</u>: <u>For</u> P <u>ramified</u>, <u>a local unit</u> u <u>is a local norm</u>, ie. $(u,\sigma)_P = +1$ <u>if and only if</u> $u_1$ <u>is a square in the residue field</u>. (characteristic $\neq 2$)

Proof: If $u_1$ is a square in the residue field, we may factor the polynomial $t^2 - u_1 = f(t)$ in the residue field as $(t + \sqrt{u_1})(t - \sqrt{u_1})$.

We are assuming the characteristic of the residue field is not 2, so these two factors are relatively prime. Hence, we may apply Hensel's Lemma, and conclude that $t^2 - u$ factors in the completion $\tilde{F}(P)$. Thus $u$ has a square root in $\tilde{F}(P)$, and $(u,\sigma)_P = +1$.

Conversely, if $(u,\sigma)_P = +1$, then $u$ is a norm from $\tilde{E}(P)$, say $x\bar{x} = u$. We write $x$ as $x = w\pi^r$, for $w$ a local unit, $\pi$ a local uniformizer. Then $x\bar{x} = w\bar{w}(\pi^r\bar{\pi}^r) = u$. Since $u$ is a local unit, $r = 0$. Thus $u$ is a square in $\tilde{O}_E(P)/\tilde{m}(P)$, since the induced involution is trivial there. However, $\tilde{O}_E(P)/\tilde{m}(P) = O_E(P)/m(P)$, so that $u_1$ is a square in the residue field.

2.5 If $P_\infty$ is infinite ramified: $(a,\sigma)_\infty = -1$ if and only if $a < 0$ and $\sigma < 0$. This is clear as the completion with respect to $P_\infty$ is $\mathbb{R}$.

We restate the Theorem of Hasse in terms of symbols.

Theorem 2.6 $y \in F^*$ is a norm from $E^*$ if and only if $(y,\sigma)_P = +1$ at all primes $P$, finite and infinite.

We also list the important properties of the Hilbert symbols, in addition to the discussion above.

2.7 $(a,\sigma)_P = +1$ for almost all $P$ since at almost all $P$, $a$ and $\sigma$ are both units [O'M 166]. Almost all means all but finitely many in this case.

2.8 Hilbert Reciprocity $\prod_P (a,\sigma)_P = +1$

2.9 Realization: If $\varepsilon(P) \in Z^*$ is a function defined for all $P$ satisfying

(1) $\varepsilon(P) = +1$ if $P$ splits

(2) $\varepsilon(P) = +1$ at all but finitely many primes

(3) $\prod_P \varepsilon(P) = +1$

then there is an $f \in F^*$ with $(f,\sigma)_P = \varepsilon(P)$.

We again refer to O'Meara [O'M 203].

Note: At non-split primes, $n_P = 2$, and $\sigma$ is not a square in $\tilde{F}(P)$.

3. Rank

Let $[M,B] \in H_u(E)$. We define the rank mod 2 of $[M,B]$, denoted $rk[M,B]$, by

$$rk[M,B] = 0 \text{ if } [M:E] \text{ is even.}$$
$$= 1 \text{ if } [M:E] \text{ is odd.}$$

Here $[M:E]$ is the rank of the vector space $M$ over $E$.

<u>Theorem 3.1</u> rk: $H_u(E) \to F_2$ <u>is a well-defined group homomorphism.</u>

<u>Proof</u>: The only problem is to show that rk is well-defined. So, let $[M,B] \in H_u(E)$ have $[M,B] = 0$. Then there is a metabolizer $N \subset M$ with $N = N^{\perp}$. This yields the exact sequence:

$$0 \to N^{\perp} \to M \overset{Ad_R B}{\to} \mathrm{Hom}_E(N,E) \to 0$$

$$\begin{aligned} \text{Hence} \quad \mathrm{rank}\, M &= \mathrm{rank}\, N^{\perp} + \mathrm{rank}(\mathrm{Hom}_E(N,E)) \\ &= \mathrm{rank}\, N^{\perp} + \mathrm{rank}\, N \\ &= \mathrm{rank}\, N + \mathrm{rank}\, N \\ &= 2\mathrm{rank}\, N. \end{aligned}$$

Thus $[M,B] = 0$ implies $\mathrm{rk}[M,B] = 0$. It follows that rk is well-defined.

Clearly rk is additive, so rk defines a group homomorphism. We also note that rk is in fact a ring homomorphism. □

<u>Corollary 3.2</u> rk: $H_u(I) \to F_2$ <u>defined as above is a well-defined group homomorphism. Here</u> $H_u(I)$ <u>denotes</u> I-<u>valued</u> u <u>Hermitian inner products on torsion free</u> D-<u>modules.</u>

<u>Proof</u>: Apply I 5.4 . □

## 4. <u>Diagonalization and the discriminant</u>

In order to discuss the discriminant, we must establish some notation. We first pick a fixed basis, $\{e_1,\ldots,e_n\}$ for M .

Thus, if $x \in M$, we write $x = (a_1,\ldots,a_n)$ to mean $x = \sum_{i=1}^{n} a_i e_i$ , $a_i \in E$.

Associated to the inner product $B: M \times M \to E$, there is the matrix $B' = (b_{ij})$, where $b_{ij} = B(e_i,e_j)$. If $x = (a_1,\ldots,a_n)$, and $y = (b_1,\ldots,b_n)$, then in terms of $B'$ we have

$$(a_1,\ldots,a_n)B'\begin{pmatrix}\bar{b}_1\\ \vdots\\ \bar{b}_n\end{pmatrix} = B(x,y)$$

which we also write as $B(x,y) = xB'\bar{y}^t$ . This follows since $B$ is linear in the first variable and conjugate linear in the second.

Now $[M,B] \in H_u(E)$. Thus $B(e_i,e_j) = u\overline{B(e_j,e_i)}$ so that $b_{ij} = u\bar{b}_{ji}$ . It follows that $B'$ satisfies $B' = u\bar{B}'^t$.

We now let $\{e_i^\#\}_{i=1}^{n}$ denote the dual basis to $\{e_i\}$ . $e_i^\# : M \to E$ is defined on a basis of $M$ by $e_i^\# (e_j) = \delta_{ij}$, the Kronecker $\delta$ , and extended linearly to $M$.

We consider the adjoint map of $B$, $Ad_R B: M \to Hom_E(M,E)$. $Ad_R B: e_i \to B(-,e_i)$. We express $(Ad_R B)(e_i)$ as a linear combination of the $\{e_j^\#\}$ . This yields

$$(Ad_R B)e_i = \sum_{j=1}^{n} B(e_j,e_i)e_j^\# .$$

We thus see that the matrix of the adjoint transformation, in terms of the bases $\{e_i\}$, $\{e_i^\#\}$ is none other than $(b_{ij}) = B' = (B(e_i,e_j))$. We can thus state:

<u>Proposition 4.1</u> <u>Given a bilinear map</u> $B: M \times M \to E$, <u>the adjoint</u> $Ad_R B: M \to Hom_E(M,E)$ <u>is an isomorphism if and only if</u>

$(B(e_i,e_j))$ <u>is an invertible matrix</u>. □

Next, we wish to relate $u$ Hermitian to $1$ Hermitian.

<u>Proposition 4.2</u> $H_u(E) \simeq H_1(E)$.

<u>Proof</u>: Since $u$ satisfies $u\bar{u} = 1$, by Hilbert's Theorem 90, we can find $x_1 \in E$ with $x_1\bar{x}_1^{-1} = u$. $x_1$ then clearly yields an isomorphism: $H_u(E) \simeq H_1(E)$, merely by scaling the inner product with $x_1^{-1}$. In other words, $[M,B] \in H_u(E) \to [M,B_1] \in H_1(E)$ where $B_1(x,y) = (1/x_1)B(x,y)$. We must check $B_1$ is $1$ Hermitian. $B_1(x,y) = (1/x_1)B(x,y) = (u/x_1)\overline{B(y,x)} = (1/\bar{x}_1)\overline{B(y,x)} = \overline{B_1(y,x)}$.

Conversely, if $[M,B_1] \in H_1(E)$, we must check that B is u Hermitian, where $B(x,y) = x_1B_1(x,y)$. We compute: $B(x,y) = x_1B_1(x,y) = u\bar{x}_1B_1(x,y) = u\bar{x}_1\overline{B_1(y,x)} = u(\overline{x_1B_1(y,x)}) = u\overline{B(x,y)}$. □

<u>Remark</u>: We could choose $x_1 = 1 + u$.

When the characteristic of $E$ is not 2, we shall see that it is possible to choose a basis for $M$ so that the matrix $B'$ of B is diagonalized. We prove this first for $[M,B] \in H_1(E)$. For $[M,B] \in H_u(E)$, we apply Proposition 4.2 and the above, observing that the isomorphism given in 4.2 preserves diagonalization.

Now let $[M,B] \in H_1(E)$. By the trace lemma, $B_1$ defined on M by $B_1(x,y) = tr_{E/F} \circ B(x,y)$ is a non-singular symmetric bilinear form on M. Here $tr_{E/f}$ denotes the trace map. Since $B_1(x,y) = (1/2)\big(B_1(x + y,x + y) - B_1(x,x) - B_1(y,y)\big)$, and $B_1 \neq 0$, it follows that there exists $v \in M$ with $B_1(v,v) \neq 0$. It follows that $B(v,v) \neq 0$ also. Extend $v$ to a basis of $M$, $\{v,v_2,\ldots,v_n\}$.

Notice that $\{v, v_2 - (B(v_2,v)/B(v,v))v, \ldots, v_n - (B(v_n,v)/B(v,v))v\} = \{w_i\}$ is also a basis for M. The computations:

$$B(v_i - (B(v_i,v)/B(v,v))v,v) = B(v_i,v) - B(v_i,v) = 0$$
$$= \overline{B(v,v_i - (B(v_i,v)/B(v,v))v)}$$

show that with respect to $\{w_i\}$, the matrix of B looks like:

$$\begin{pmatrix} B(v,v) & 0 & \cdots & 0 \\ 0 & & & \\ \vdots & & & \\ 0 & & & \end{pmatrix}.$$

Continuing, consider $B_1(x,y)$, for y in the span of $\{w_2,\ldots,w_n\}$, $x \in M$. Again, since $B_1$ is non-singular, we can find x with $0 \neq B_1(x,y) = (1/2)(B_1(x+y,x+y) - B_1(x,x) - B_1(y,y))$. Write $x = \sum_{i=1}^{n} a_i w_i$. Then it is clear that:

$$B_1(x,y) = B_1(\sum_{i=2}^{n} a_i w_i, y) + B_1(a_1 w_1, y) = B_1(\sum_{i=2}^{n} a_i w_i, y).$$

Thus, we can find $v \in \langle w_2,\ldots,w_n \rangle$ with $B(v,v) \neq 0$. Continuing inductively, we form $\{w_1, v, \ldots\}, \ldots,$ and diagonalize B. We may thus state:

<u>Proposition 4.3</u> <u>Given</u> $[V,B] \in H_1(E)$, <u>there is a basis for</u> V <u>which makes the matrix of</u> B <u>diagonal</u>. (characteristic of $E \neq 2$) □

Remark 4.4 This also holds for $H_u(E)$ by applying 4.2.

Remark 4.5 As long as the involution $-$, on $E$ is non-trivial, we may prove 4.3 directly even if the characteristic of $E$ is 2. In order to see this, we must show how to produce a vector $v$ with $B(v,v) \neq 0$. Suppose to the contrary that $B(v,v) = 0$ for all $v \in M$. Assuming that $B$ is non-singular, so not identically 0, we can find $v, w \in M$ with $B(v,w) \neq 0$. However $B(v+w,v+w) = 0 = B(v,w) + B(w,v)$. Thus $B(v,w) = -B(w,v)$. Hence, for any $a \in E$,

$$aB(v,w) = B(av,w) = -B(w,av) = -\bar{a}B(w,v)$$

Since $B(v,w) \neq 0$, $a = \bar{a}$ for all $a \in E$, and the involution on $E$ is trivial. Contradiction.

Once we have such a vector $v$, we proceed as in 4.3 to produce an orthogonal basis.

Remark 4.6 Thus, we see that 4.3 holds for $H_u(E)$, provided we are not in the situation of a trivial involution or a field of characterisitc 2. For $[M,B] \in H_u(E)$, where the characteristic of $E$ is 2, we may write $B$ as a direct sum of 1-dimensional forms and metabolic forms,

$$\begin{pmatrix} 0 & 1 \\ -1 & 0 \end{pmatrix},$$

see [K-1 22].

Diagonalizing an inner product space $(M,B)$ means choosing a basis of $M$ with respect to which the matrix of $B$ is diagonal. In other words, $(M,B) \simeq \oplus (M_i,B_i)$, where $M_i$ is a 1-dimensional vector space. It is natural then to compare the matrices of $B$ given by different choices of bases for $M$.

Suppose that $\{e_1,\ldots,e_n\}$ and $\{f_1,\ldots,f_n\}$ are two bases of $M$. We write $E$ = matrix of $B$ with respect to $\{e_i\}$, and $F$ = matrix of $B$ with respect to $\{f_j\}$.

We may express $\{e_i\}$ in terms of $\{f_j\}$. Suppose $e_i = \sum_{j=1}^{n} c_{ij}f_j$, and let $C = (c_{ij})$, $C^t$ = transpose of $C$.

<u>Proposition 4.7</u> $E = CF\bar{C}^t$

<u>Proof</u>: In terms of $\{e_i\}$, write $e_i = (0,\ldots,\underset{i^{th}\text{ place}}{1},\ldots,0)$, $e_j = (0,\ldots,1,\ldots,0)$, and $e_iE\bar{e}_j^t = e_{ij} = B(e_i,e_j)$. The ij component of $CF\bar{C}^t$ is likewise given by:

$$\underset{i^{th}\text{ place}}{(0,\ldots,1,\ldots,0)}CF\bar{C}^t\begin{pmatrix}0\\ \vdots\\ 1\\ \vdots\\ 0\end{pmatrix} j^{th}\text{ place}$$

$$= (c_{i1},\ldots,c_{in})F\begin{pmatrix}\bar{c}_{j1}\\ \vdots\\ \bar{c}_{jn}\end{pmatrix} = B(c_{i1}f_1 + \quad + c_{in}f_n, \; c_{j1}f_1 + \quad + c_{jn}f_n)$$

$$= B(e_i,e_j)$$

$$= e_{ij} \quad \text{as above.} \quad \square$$

We would now like a Witt group invariant corresponding to the determinant of a matrix. This invariant should be independent of the choice of basis, as well as the Witt representative of the given

Witt equivalence class.

Let $[M,B] \in H_1(E)$. Let $B_1$ and $B_2$ denote two different matrices of $B$. By 4.7, we can write $B_1 = CB_2\bar{C}^t$, for a non-singular matrix $C$. Let $\det B_1$ denote the determinant of $B_1$. Then

$$\begin{aligned} \det B_1 &= \det C \ \det B_2 \ \det \bar{C} \\ &= \det C \ \det B_2 \ \overline{(\det C)} \ . \end{aligned}$$

Thus, we can read the determinant of $B$ in $F^*/NE^*$, since when $B_1$ is diagonalized, the diagonal elements must be in $F^*$ as $B$ is Hermitian. Unfortunately, this is not a Witt invariant. For example, $\det B$ need not be in $NE^*$ even when $[V,B] = 0$ as we see below.

$$B = \begin{pmatrix} 0 & 1 \\ 1 & 0 \end{pmatrix} \ . \quad \det B = -1,$$

which may not be a norm.

We are thus led to define a corresponding notion:

<u>Definition 4.8</u> <u>Let</u> $B_1$ <u>be a matrix with coefficients in</u> $E$, <u>corresponding to a Hermitian form</u> $B$. <u>Then</u> $\text{dis } B = (-1)^{n(n-1)/2}\det B_1$, <u>where</u> $n$ <u>is the dimension of</u> $M$, <u>is called the discriminant of the inner product space</u> $(M,B)$.

<u>Lemma 4.9</u> <u>If</u> $[M,B] = 0$ <u>then</u> $\text{dis } B \in NE^*$.

<u>Proof</u>: Let $N$ be a metabolizer for $M$. Let $\{n_1,\dots,n_t\}$

be a basis for N. Extend this to a basis for M, say $\{n_1,\ldots,n_t,n_{t+1},\ldots,n_{2t}\}$ . With respect to this basis, B has matrix

$$\begin{pmatrix} 0 & C \\ \bar{C} & X \end{pmatrix}.$$

Interchanging the first t colums with the last t columns, we obtain a matrix

$$\begin{matrix} C & 0 \\ X & \bar{C} \end{matrix}.$$

This requires interchanging $t^2$ columns. Hence B has

$$\det B = (-1)^{t^2} \det C \cdot \det \bar{C} = (-1)^t \det C \cdot \det \bar{C}.$$

Thus,

$$\begin{aligned} \text{dis } B &= (-1)^{2t(2t-1)/2}(-1)^t \det C \cdot \det \bar{C} \\ &= (-1)^{t+t(2t-1)} \det C \cdot \det \bar{C} \\ &= (-1)^{t+t} \det C \cdot \det \bar{C} \\ &= \det C \cdot \det \bar{C} \in NE^* \end{aligned}$$

as claimed. □

It follows that dis is exactly the kind of invariant we seek. There is still a problem, namely dis is not additive. Hence, we do not obtain a group homomorphism:

$$H_1(E) \rightarrow F^*/NE^*.$$

To remedy this problem we invent the group

$$Q(E) = F^*/NE^* \times F_2 \qquad F_2 = \{0,1\}$$

[Lm 38]. The binary operation in $Q(E)$ is given by:

$$(d_1,e_1) \cdot (d_2,e_2) = ((-1)^{e_1 e_2} d_1 d_2, e_1 + e_2)$$

This is an associative, commutative operation with (1,0) the additive identity. The additive inverse of $(d,e)$ is $((-1)^e d,e)$.

In fact, $Q(E)$ becomes a ring when one defines multiplication by:

$$(d_1,e_1) \odot (d_2,e_2) = (d_1^{e_2} d_2^{e_1}, e_1 e_2)$$

The multiplicative identity is (1,1).

<u>Proposition 4.10</u> <u>The map</u> $\phi$: $H_1(E) \to Q(E)$ <u>defined by</u> $[M,B] \to (\text{dis } B, \text{rk } M)$ <u>is a group homomorphism</u>.

<u>Proof</u>: Consider $[M,B]$ and $[W,B_1]$ in $H_1(E)$. Suppose that rank $M = n$ and rank $W = m$. We have then:

$$\phi([M,B]) = ((-1)^{n(n-1)/2}\det B,n)$$

$$\phi([W,B_1]) = ((-1)^{m(m-1)/2}\det B_1,m)$$

$$\phi([M,B]) \cdot \phi([W,B_1]) = ((-1)^{nm}(-1)^{n(n-1)/2}(-1)^{m(m-1)/2} \det B \det B_1, n+m)$$

$$\phi([M,B]+[W,B_1]) = ((-1)^{(n+m)(n+m-1)/2} \det B \det B_1, n+m)$$

But

$$(-1)^{nm+n(n-1)/2 + m(m-1)/2} = (-1)^{(n^2-n+m^2-m+2nm)/2}$$
$$= (-1)^{((n+m)^2-(n+m))/2}$$
$$= (-1)^{(n+m)(n+m-1)/2}$$

Thus $\phi$ indeed gives a well-defined group homomorphism. □

As an exercise, the reader should verify that $\phi$ is actually a ring homomorphism.

We next consider the kernel of the rank homomorphism rk, which we shall call $J$. Thus $J$ is the subgroup of $H_1(E)$ generated by the even dimensional forms.

**Proposition 4.11** $H_1(E)/J^2$ *is isomorphic to* $Q(E)$.

*Proof:* $J$ is additively generated by 2-dimensional forms, $<1,a>$. To see this, write $<a,b> \sim <1,a> - <1,-b>$. Thus $J^2$ is additively generated by the forms

$$<1,a> \otimes <1,b> = <1,a,b,ab>.$$

Applying $\phi$ to a generator, we obtain:

$$\phi<1,a,b,ab> = ((-1)^6(ab)^2,0) = (1,0).$$

Thus, $\phi$ induces a map $\tilde{\phi}$ : $H_1(E)/J^2 \to Q(E)$. We now construct an inverse of $\tilde{\phi}$ , $\gamma$ .

Define $\gamma: \ Q(E) \to H_1(E)/J^2$ by

$(a,0) \to <1,-a>$ modulo $J^2$

$(a,1) \to <a>$ modulo $J^2$ .

It is easy to check that $\gamma$ is a homomorphism, and $\gamma \circ \phi = \mathrm{id}$ $\phi \circ \gamma = \mathrm{id}$, where id is the appropriate identity map. □

Proposition 4.11 implies that $\tilde{\phi}$ is 1-1. Hence,

__Corollary 4.12__ $J^2$ __consists of even dimensional forms__ $[M,B]$, __with__ $\mathrm{dis}\, B = 1 \in F^*/NE^*$, ie. $\det B = (-1)^{n(n-1)/2}$ __where__ $n = \mathrm{rank}\ m$. □

__Corollary 4.13__ __Restricting__ $\tilde{\phi}$ __to__ $J/J^2$, __we have__ $J/J^2 \approx F^*/NE^*$. □

In fact, we may think of this as follows: $F^*/NE^*$ is embedded into $Q(E)$ by: $d \to (d,0)$. This is a subgroup of index 2. We may represent the non-identity coset by $(1,1)$. $(1,1)^2 = (-1,0)$. Thus, we have the exact sequence:

$$1 \to F^*/NE^* \to Q(E) \xrightarrow{q_2} F_2 \to 0$$

$q_2$ is projection onto the second factor. By the above remarks.

__Corollary 4.14__ __This sequence splits if and only if__ $(1,0) = (-1,0)$ __in__ $Q(E)$ __if and only if__ $-1$ __is a norm in__ $F^*/NE^*$. □

__Remark 4.15__ This defines the discriminant for $H_1(E)$. In order to define discriminant for $H_u(E)$, we fix an isomorphism:

$$f_{x_1}:\ H_u(E) \to H_1(E), \text{ by Proposition 4.2}$$
$$B \to B_1 \quad B_1(x,y) = (1/x_1)B(x,y).$$

Then define dis B = dis $f_{x_1}B$ = dis $B_1$. We must note that this depends on the isomorphism chosen ie. this depends on $x_1$, where $x_1\bar{x}_1^{-1} = u$.

Remark 4.16 The discriminant inner product space, Chapter I 3.5, yields the information crucial to the discriminant invariant above, namely the determinant of B. Its advantage is that it generalizes the notion to H(D), for D the Dedekind ring of integers.

5. Signatures

The real infinite primes, $P_\infty$, give rise to the signature invariant which we now discuss. Suppose then that $E = F(\sqrt{\sigma})$, and $\sigma < 0$ with respect to $P_\infty$. Thus $P_\infty$ is an infinite ramified prime.

Lemma 5.1 If $x \in E$, then $N_{E/F}(x) > 0$ with respect to $P_\infty$.

Proof: $N=N_{E/F}$ denotes the norm. Write $x = a + b\sqrt{\sigma}$. Then $N(x) = a^2 - b^2\sigma > 0$ since $\sigma < 0$. □

Let $[M,B] \in H_1(E)$. By 4.3, we can find a basis $\{e_i\}$ of M in which B is diagonalized. We can thus write $M = X^+ \oplus X^-$, where $B(e_i,e_i) > 0$ for $e_i \in X^+$, $B(e_i,e_i) < 0$ for $e_i \in X^-$.

Now, let $v \in X^+$, so $v = \sum_i a_ie_i$. We compute

$$\begin{aligned} B(v,v) &= B(\Sigma a_ie_i,\Sigma a_ie_i) = \Sigma B(a_ie_i,a_ie_i) \\ &= \Sigma a_i\bar{a}_iB(e_i,e_i) = \Sigma N(a_i)B(e_i,e_i) > 0 \end{aligned}$$

by Lemma 5.1.

Similarly, for all $v \varepsilon X^-$, $B(v,v) < 0$. We now define:

$$\text{sgn}[M,B] = \dim X^+ - \dim X^- .$$

sgn[M,B] is called the signature of [M,B]. In order to show sgn is well-defined, we first need:

<u>Lemma 5.2</u> sgn[M,B] <u>is independent of the basis chosen for</u> M.

<u>Proof</u>: Suppose M has two bases, $\{e_i\}$, $\{f_i\}$ which make B diagonal. Let $B_1$, $B_2$ be the matrices of B with respect to $\{e_i\}$, $\{f_i\}$.

Consider $[M,B_1] - [M,B_2]$ which is Witt equivalent to 0. With respect to the basis $\{e_i,f_i\}$ of $M \oplus M$, $B_1 \oplus -B_2$ has matrix

$$\begin{pmatrix} B_1 & 0 \\ 0 & -B_2 \end{pmatrix} .$$

It follows that $\text{sgn}[M \oplus M, B_1 \oplus -B_2] = \text{sgn } B_1 - \text{sgn } B_2$. Thus in order to show $\text{sgn } B_1 = \text{sgn } B_2$, it clearly is sufficient to show: Any metabolic space [V,h] has sgn[V,h] = 0 with respect to an arbitrary basis.

So suppose $V = X^+ \oplus X^-$. Let N be a metabolizer for V, $N = N^{\perp}$. We note that $n \varepsilon N$ implies $h(n,n) = 0$. Thus, by the remarks preceding this theorem, $X^+ \cap N = 0 = X^- \cap N$. However, $X^+ \cap N = 0$ implies $\dim N \leq \dim V - \dim X^+ = \dim X^-$. Similarly, $\dim N \leq \dim X^+$. Now, $2\dim N = \dim V$, so that $\dim X^-$ and $\dim X^+$ are both $\geq (1/2)\dim V$. However, $\dim X^+ + \dim X^- = \dim V$.

Thus, dim $X^+$ = dim $X^-$ = (1/2)dim V, and sgn[V,h] = 0 with respect to any basis. □

In the process of this proof, we have shown:

<u>Corollary 5.3</u> <u>If</u> [V,h] <u>is metabolic</u>, sgn[V,h] = 0. □

It is thus clear that sgn gives a well-defined Witt-invariant, which is a group homomorphism:

$$\text{sgn}: H_1(E) \to Z.$$

It is clear that if [M,B] has finite order, sgn[M,B] = 0, since every element in Z has infinite order. Thus sgn is non-trivial only on the non-torsion elements in $H_1(E)$.

We finally recall Landherr's Theorem which explicitly computes $H_1(E)$, for E an algebraic number field [Lh] .

<u>Landherr's Theorem 5.4</u> <u>There is an exact sequence</u>:

$$0 \to (4Z)^r \to H(E) \overset{\phi}{\to} Q(E) \to 0$$

$\phi$[M,B] = (dis B, rk V).

<u>The kernel of</u> $\phi$, ker$\phi$ , <u>is determined by the real infinite ramified primes and the corresponding signatures, each of which is divisible by</u> 4. <u>Here</u> r <u>is the number of real infinite primes. When</u> r = 0, $\phi$ <u>is an isomorphism</u>.

This theorem is important in the boundary computation in Chapter VI.

By Proposition 4.10, $H(E)/J^2 \simeq Q(Z)$, so we can state:

Corollary 5.5 $J^2 \simeq 4(\underbrace{Z \oplus Z \ \ldots \oplus Z}_{r \text{ times}})$ □

Remark 5.6 As with the discriminant, we can define a signature invariant for $H_u(E)$. This is done by picking an isomorphism $f_{x_1}: H_u(E) \to H_1(E)$ as in 4.15. We then define

$$\operatorname{sgn}[M,B] = \operatorname{sgn}(f_{x_1}[M,B]) .$$

Chapter III POLYNOMIALS

Given a Witt equivalence class, $[M,B,\ell]$ in $W(k,F)$, we shall decompose it as $[M,B,\ell] = \oplus[M_i,B_i,\ell_i]$, according to the irreducible factors of the characteristic polynomial of $\ell$. This is the object in Chapter IV.

In this chapter, we lay the groundwork for the above decomposition. This involves a careful study of the characteristic and minimal polynomials of $\ell$. These polynomials belong to $K(F) = \{p(t): p(t)$ is a monic polynomial with non-zero constant term, coefficients in $F$ a field$\}$. We assume throughout this section that we are working over a field $F$.

On $K(F)$ we define an involution $T_k: K(F) \to K(F)$. The characteristic and minimal polynomials are shown to be $T_k$ fixed.

When $p(t)$ is irreducible and $T_k$ fixed, we consider the field $F[t,t^{-1}]/(p(t)) = F(\theta)$. It is shown that there is an induced involution of $F(\theta)$ given by $\bar{\theta} = k\theta^{-1}$.

This discussion provides the key ingredients for the computations to be made later.

Let $(M,B)$ be an inner product space, and let $\ell: M \to M$ be $F$-linear. Recall the adjoint, $\ell^*$ of $\ell$ is defined by the equation $B(v,\ell w) = B(\ell^* v,w)$ [I 4.9].

<u>Lemma 1.1</u> $\ell$ <u>and</u> $\ell^*$ <u>have the same characteristic polynomial and the same minimal polynomials</u>.

<u>Proof</u>: For any polynomial $p(t)$, $B(p(\ell^*)v,w) = B(v,p(\ell)w)$.

Thus, $p(\ell^*) = 0$ if and only if $p(\ell) = 0$, since $B$ is non-singular. The assertion about minimal polynomials follows.

Working over a field, we may view $M$ as the space of $n \times 1$ column matrices and $B$ as an $n \times n$ matrix $B'$, [II 4]. $B(v,w) = v^t B' \bar{w}$, where $v^t$ denotes the transpose of $v$, $\bar{w}$ denotes the conjugate of $w$,

$$w = \begin{pmatrix} a_1 \\ \vdots \\ a_n \end{pmatrix} \quad , \quad \bar{w} = \begin{pmatrix} \bar{a}_1 \\ \vdots \\ \bar{a}_n \end{pmatrix} .$$

$\ell$ is multiplication by an $n \times n$ matrix $L$. To simplify our notation, we identify $B$ with its matrix and write $B' = B$.

We compute, $[(B^{-1}L^tB)v]^t B\bar{w} = v^t BLB^{-1}B\bar{w} = v^tB(L\bar{w})$. It follows that $L^* = B^{-1}L^tB$ = matrix of $\ell^*$. Hence, letting det denote the determinant, and $I$ the $n \times n$ identity matrix,

$$\begin{aligned} \det(tI - B^{-1}L^tB) &= (\det B)(\det (tI - B^{-1}L^tB))(\det B^{-1}) \\ &= \det (BtB^{-1} - BB^{-1}L^tBB^{-1}) \\ &= \det (tI - L^t) \\ &= \det (tI - L). \end{aligned}$$

The assertion for characteristic polynomials follows. □

Let $\ell$ be a map of degree $k$. Then $\ell$ is non-singular, and $\ell$ is related to $\ell^*$ by:

<u>Lemma 1.2</u> *If $\ell$ has matrix $L$, and $\ell^*$ has matrix $L^*$, then $L^* = kL^{-1}$.*

<u>Proof</u>: $B(\ell^* v,w) = B(v,\ell w) = B(\ell\ell^{-1}v,\ell w) = kB(\ell^{-1}v,w) = B(k\ell^{-1}v,w)$. Again, since $B$ is non-singular, it follows that $\ell^* = k\ell^{-1}$. □

<u>Proposition 1.3</u> <u>Let</u> $\ell$ <u>be a map of degree</u> $k$. <u>Then both the minimal and characteristic polynomials of</u> $\ell$ <u>satisfy</u>:

$$t^{\text{degree } p(t)} p(t^{-1}k) = a_0 p(t) ,$$

<u>where</u> $a_0$ = <u>constant term of</u> $p(t)$ .

<u>Proof</u>: Let $\chi(t)$ = characteristic polynomial of $\ell$ . Since $\ell$ is non-singular the constant term of $\chi(t)$ is non-zero. (Of course the dimension of the vector space $M$ is $n$, the degree of $\chi(t)$).

$\ell^* = k\,\ell^{-1}$ by Lemma 1.2. Thus, by Lemma 1.1, $\ell$ and $k\ell^{-1}$ have the same characteristic polynomial. The identity

$$(-t^{-1}L)(tI-kL^{-1}) = (kt^{-1}I - L)$$

yields

$$\det(-t^{-1}L)\chi(t) = \chi(kt^{-1}) .$$

However,

$$\begin{aligned} \det(-t^{-1}L) &= (-1)^n t^{-n} \det L \\ &= (-1)^n t^{-n} (-1)^n a_0 \\ &= t^{-n} a_0 \quad , \quad a_0 = \text{constant term of } \chi(t). \end{aligned}$$

Thus, $t^{-n}a_0\chi(t) = \chi(kt^{-1})$, so that $t^n\chi(kt^{-1}) = a_0\chi(t)$ as desired.

In order to check the result for the minimal polynomial $p(t)$ of $\ell$ we again use $\ell^* = k\ell^{-1}$. Let degree $p(t) = m$. By Lemma 1.1

$p(k\ell^{-1}) = 0$. Thus, $q(t) = a_0^{-1}t^m p(kt^{-1})$ is a monic polynomial of degree $m$ = degree $p(t)$ with $q(\ell) = 0$. Hence $p(t) = a_0^{-1}t^m p(kt^{-1})$ as claimed. □

We continue the study of these polynomials by letting $K(F) = \{p(t): p(t)$ is a monic polynomial, with constant term $a_0 \neq 0\}$.

Here, $p(t) = \sum_{i=0}^{n} a_i t^i$. This is a cancellation semigroup with respect to multiplication of polynomials. Further, any polynomial in $K(F)$ can be uniquely factored into a product of powers of irreducible polynomials in $K(F)$.

For $k \neq 0$, $k \in F^*$, we are led by Proposition 1.3 to introduce an automorphism of period 2 on $K(F)$ by:

$$T_k: \ p(t) \to t^{\deg p(t)} a_0^{-1} p(kt^{-1}) = (T_k p)(t).$$

Proposition 1.3 then says that for a degree k mapping structure $(M,B,\ell)$, both the characteristic and minimal polynomial of $\ell$ are $T_k$ fixed.

<u>Lemma 1.4</u> <u>A polynomial</u> $p(t)$ <u>is fixed under</u> $T_k$ <u>if and only if its coefficients satisfy</u> $a_j k^j = a_0 a_{n-j}$, $0 \leq j \leq n$ = degree $p(t)$.

<u>Proof</u>: Clear by definition of $T_k$. □

We note that $a_0^2 = k^n$.

To summarize, if $p(t) \in K(F)$ is $T_k$ fixed, exactly one of the following three cases applies.

Type 1: deg p(t) = 2n and $a_0 = k^n$.

Thus

$$a_j = k^{n-j} a_{2n-j} \qquad 0 \le j \le n .$$

Type 2: deg p(t) = 2n and $a_0 = -k^n$. Assume char $F \neq 2$.

Thus

$$a_j = -k^{n-j} a_{2n-j} \qquad 0 \le j \le n ,$$

so that

$$a_n = -a_n = 0 .$$

Note: There is no loss of generality in assuming characteristic $F \neq 2$ in this case.

Type 3: deg p(t) = 2d + 1.

Thus

$$k^{2d+1} = a_0{}^2 , \text{ and } k = (a_0/k^d)^2 .$$

Lemma 1.5 *If* $p(t) \in K(F)$ *is* $T_k$ *fixed, of degree* 2d + 1, *then* $-a_0 k^{-d}$ *is a root of* p(t).

Proof: Consider $p(-a_0 k^{-d})$. The 2j term is $a_{2j}(-a_0 k^{-d})^{2j} = a_{2j} k^j$. However, this 2j term cancels with the 2(d - j) + 1 term since

$$a_{2(d-j)+1}(-a_0 k^{-d})^{2(d-j)+1} = a_{2(d-j)+1}(-a_0 k^{-d})(-a_0 k^{-d})^{2(d-j)}$$

However,

$$a_{2(d-j)+1}a_0 = a_{2j}k^{2j}, \text{ and } (-a_0k^{-d})^2 = k.$$

So the above equals:

$$= -a_{2j}k^{2j}k^{-d}k^{d-j} = -a_{2j}k^j \qquad \square$$

<u>Lemma 1.6</u> *If* $p(t) \varepsilon K(F)$ *is of type* 2, *then* $(t^2 - k)$ *divides* $p(t)$. (Characteristic $F \neq 2$)

<u>Proof</u>: For $0 \leq j < n$, $p(\sqrt{k})$ will have $j^{th}$ term $a_j(\sqrt{k})^j$, and $(2n - j)^{th}$ term $a_{2n-j}(\sqrt{k})^{2n-j}$. Further,

$$\begin{aligned} a_j(\sqrt{k})^j &= -k^{n-j}a_{2n-j}(\sqrt{k})^j \\ &= -(\sqrt{k})^{2(n-j)+j}a_{2n-j} \\ &= -a_{2n-j}(\sqrt{k})^{2n-j} \end{aligned},$$

and these terms cancel.

Since char $F \neq 2$, $a_n = 0$ and $\sqrt{k}$ is a root of $p(t)$. Hence we can write $(t^2 - k)q(t) = p(t)$ over $F(\sqrt{k})$. It is clear that $q(t) \varepsilon F[t]$, since $t^2 - k$ and $p(t)$ are. $\square$

Hence, irreducible polynomials in $K(F)$ which are $T_k$ fixed fall into the three following types.

<u>Type 1</u>: deg $p(t) = 2n$ and $a_0 = k^n$

Type 2: $k \notin F^{**}$ and $t^2 - k = p(t)$, when char $F \neq 2$

Type 3: $k \in F^{**}$ and $p(t) = t \pm \sqrt{k}$

On $F[t,t^{-1}]$ we introduce the involution: $t \to kt^{-1}$, $t^{-1} \to k^{-1}t$. Denote this by: $\gamma \to \bar{\gamma}$. See [VI 2].

Let $\gamma \in F[t,t^{-1}]$, say $\gamma = \sum_{-m}^{n} A_j t^j$. Then $\gamma = \bar{\gamma}$ if and only if $n = m$ and we have $A_{-j} = A_j k^j$, $0 \leq j \leq n$.

Suppose, $\gamma = \bar{\gamma}$ and $A_n = 1$. Then $t^n\gamma = p(t)$ belongs to $K(F)$ and is a $T_k$ fixed polynomial of type 1. Conversely, any $T_k$ fixed polynomial of type 1 can be written as $t^n\gamma = p(t)$, for a unique $\gamma = \bar{\gamma}$, where $2n$ = degree $p(t)$.

Continuing, suppose $p(t)$ is a $T_k$ fixed polynomial of type 2. Then by Lemma 1.6, $p(t) = (t^2 - k)q(t)$. However, $T_k$ is multiplicative, and $t^2 - k$ is $T_k$ fixed. It follows that $q(t)$ is also a $T_k$ fixed polynomial. $q(t)$ has degree $2(n - 1)$ and constant term $k^{n-1}$.

Hence $p(t) = (t^2 - k)q(t)$ where $q(t)$ is a $T_k$ fixed polynomial of type 1, or $q(t) = 1$. So we can write $p(t) = (t^2 - k)t^{n-1}\gamma = t^n(t - kt^{-1})\gamma$ where $\gamma = \bar{\gamma}$ and degree $p(t) = 2n$.

Finally for type 3, let $p(t)$ have constant term $a_0$. By Lemma 1.5, $p(t) = (t + a_0k^{-d})q(t)$. As above we show $q(t)$ is a type 1 $T_k$ fixed polynomial of degree $2d$ or $q(t) = 1$.

**Lemma 1.7** *If $p(t) \in K(F)$ is $T_k$ fixed, then the principal ideal $(p(t)) \subset F[t,t^{-1}]$ is - invariant.*

*Proof*: We take first the case when $p(t)$ is of type 1, say

$p(t) = t^n\gamma$ . It follows that $(p(t)) = (\gamma)$ since $t$ is a unit in $F[t,t^{-1}]$. But $\gamma = \bar{\gamma}$, so that $(p(t))$ is - invariant.

Next, let $p(t)$ be of type 2. Then by the discussion before the Lemma, $p(t)$ factors as $p(t) = t^n(t - kt^{-1})\gamma$ , where $\gamma = \bar{\gamma}$ and $\overline{(t - kt^{-1})} = -(t - kt^{-1})$. Clearly then, $(p(t)) = ((t - kt^{-1})\gamma\ )$ is - invariant.

For $p(t)$ of type 3, $p(t) = t^d(t + a_0k^{-d})\gamma$, with $\gamma = \bar{\gamma}$ . Now, we compute $\overline{(t + a_0k^{-d})} = kt^{-1} + a_0k^{-d} = (t + a_0k^{-d})(a_0k^{-d}t^{-1})$ since $(a_0/k^d)^2 = k$. However, $a_0k^{-d}t^{-1}$ is a unit in $F[t,t^{-1}]$, which again yields that $(p(t))$ is - invariant. □

We summarize this discussion. Let $p(t)$ be a $T_k$ fixed irreducible polynomial in $K(F)$. Then there are three cases to consider.

<u>Type 1</u>: $F[t,t^{-1}]/(p(t)) = F(\theta)$ is a simple algebraic extension of $F$ together with a non-trivial involution $\bar{\theta} = k\theta^{-1}$. Here $\theta$ is identified with $t$. If $\bar{t} = t$, then $kt^{-1} = t$ so that $t^2 = k$, $t^2 - k = 0$ and we are in type 2.

<u>Type 2</u>: $F[t,t^{-1}]/(p(t)) = F[t,t^{-1}]/(t^2 - k) = F(\sqrt{k})$ for the case that $k \notin F^{**}$. The induced involution is $\sqrt{k} \to k(\sqrt{k})^{-1} = \sqrt{k}$, which is trivial. Note that this is not the involution $\sqrt{k} \to -\sqrt{k}$.

<u>Type 3</u>: In this case, $k \in F^{**}$, say $f^2 = k$. The irreducible polynomial is $p(t) = t \pm f$. The field $F[t,t^{-1}]/(t \pm f) \simeq F$, by identifying $t$ with $\pm f$. The involution:

$$\pm f \to k/\pm f = (\pm f)^2/(\pm f) = \pm f$$

is trivial, and so is the extension.

Finally, in the type 1 situation when the involution $-$ is non-trivial, we wish to describe the fixed field.

<u>Lemma 1.8</u> <u>In the type 1 situation, the fixed field of</u> $(F(\theta),-)$ <u>is</u> $F(\theta + k\theta^{-1})$.

<u>Proof</u>: There is the embedding $F[x] \to F[t,t^{-1}]$ given by $x \to t + kt^{-1}$. We claim that the image of $F[x]$ is the subring of $-$ fixed elements.

Let $\gamma \in F[t,t^{-1}]$, $\gamma = \sum_{-n}^{n} A_j t^j$ with $A_{-j} = A_j k^j$, be a typical $-$ fixed element. Consider $\gamma - A_n(t + kt^{-1})^n$. This is still $-$ fixed, and can be written as $\sum_{-n+1}^{n-1} B_j t^j$. Continuing inductively,

$$= \sum_{i=0}^{n} a_i (t + kt^{-1})^i = q(t + kt^{-1}) \text{ as claimed.} \quad \square$$

Suppose $A_n = 1$ and $\gamma$ is the image of a monic polynomial, $q(x)$, in $F[x]$.

<u>Claim</u>: If $t^n\gamma = p(t)$ is irreducible, then so is $q(x)$. For if $q(x)$ factors as $q(x) = q_1(x)q_2(x)$, with $r = \text{degree } q_1(x)$, $w = \text{degree } q_2(x)$, then

$$p(t) = t^n\gamma = [t^r q_1(t + kt^{-1})][t^w q_2(t + kt^{-1})],$$

so that $p(t)$ also factors.

We may thus write $F[x]/(q(x))$ as the $-$ fixed elements in $(f(\theta),-)$. Clearly, the minimal polynomial of $\theta$ over $F(\theta + k\theta^{-1})$ is $x^2 - (\theta + k\theta^{-1})x + k$.

## Chapter IV WITT GROUP OF A FIELD

We wish to compute the Witt group $W(k,F)$ for $F$ a field. This is done by decomposing $W(k,F)$ as a direct sum of groups $W(k,F;f)$ according to the irreducible factors $f(t)$ of the characteristic polynomial of $\ell$. We identify each group,

$$W(k,F;f) \approx W(k,F;F[t]/(f(t))$$

by taking anisotropic representatives.

On $F[t]/(f(t))$ there is an induced involution by Chapter III. We prove a trace lemma which then enables us to compute these groups $W(k,F;F[t]/(f(t)))$. In this manner then we will have computed $W(k,F)$.

The trace lemma is then used in several cases to compute Witt groups. This computation is valuable for the ensuing chapters.

Finally, we prove a result showing the relation between torsion in $W(-k,F)$ and the number of squares necessary to express $k$ as a sum of squares.

### 1. Decomposition by characteristic polynomial

Given a degree k mapping structure $(M,B,\ell)$, we may view $M$ as a $D[t]$-module by defining the action of the indeterminate $t$ to be the same as $\ell$. By III.1.3, the characteristic polynomial of $\ell$, $p(t)$, is $T_k$ fixed.

__Proposition 1.1__ __If__ $(M,B,\ell)$ __is metabolic, and__ $\ell$ __has characteristic polynomial__ $p(t)$, __then__ $p(t)$ __factors as__ $p(t) = f(t) \cdot T_k f(t)$

for some monic polynomial $f(t)$.

Proof: Let $f(t)$ be the characteristic polynomial of $\ell$ restricted to $N$, where $N$ is a metabolizer for $M$.

We now make $\mathrm{Hom}_D(N,K)$ into a $D[t]$-module. This is done by defining the action of $t$ by:

Let $h \in \mathrm{Hom}_D(N,K)$. Then $(t\cdot h)(n) = h(\ell^* n)$, where $\ell^*$ is the adjoint of $\ell$. Viewed thus, $\mathrm{Ad}_R B\colon M \to \mathrm{Hom}_D(N,K)$ is a $D[t]$-module homomorphism since:

$$\begin{aligned}\mathrm{Ad}_R B(t\, m) = \mathrm{Ad}_R B(\ell m) &= B(-,\ell m) = B(\ell^*(-),m)\\ &= t\cdot B(-,m) = t\cdot \mathrm{Ad}_R B(m),\end{aligned}$$

We thus obtain an exact sequence of $D[t]$-modules:

$$0 \to N \to M \xrightarrow{\mathrm{Ad}_R B} \mathrm{Hom}_D(N,K) \to 0$$

By definition of the action of $t$ on $\mathrm{Hom}_D(N,K)$, its characteristic polynomial is simply that of $\ell^*|_N$. We can see this by identifying $N$ with its dual space, $\mathrm{Hom}_D(N,K)$. The action of $t$ on $N$ induced from the corresponding action of $t$ on $\mathrm{Hom}_D(N,K)$ above is then $\ell^*|_N$.

Note: In this section we are working over $F$ a field, so that $D = K = F$. We have used the notation $D$, $K$ to follow our previous conventions.

The question arises; what is the characteristic polynomial of $\ell^*|_N$? To begin with, by Lemma III.1.2, $\ell^* = k\ell^{-1}$ on $M$, hence all the more so on $N$. We write $L_1$ as the matrix of $\ell$ restricted to $N$. Then the matrix of $\ell^*$ is $kL_1^{-1}$. Now $f(t)$ = characteristic polynomial of $\ell = \det(tI - L_1)$. We compute

$$\begin{aligned}\det(tI - kL_1^{-1}) &= \det(-tL_1 + kI)\det(-L_1^{-1}) \\ &= \det(tI)\det(kt^{-1}I - L_1)\det(-L_1^{-1}) \\ &= t^n \det(kt^{-1}I - L_1)\det(-L_1^{-1}) \\ &= t^n f(kt^{-1})\det(-L_1^{-1}) \quad .\end{aligned}$$

where $n$ = degree $f(t)$ = dimension $N$. Here $\det(-L_1^{-1})$ is a constant; from which it follows that $\det(-L_1^{-1}) = a_0^{-1}$ where $a_0$ is the constant term of $f(t)$, and that $T_k f(t) = \det(tI - kL_1^{-1})$ = characteristic polynomial of $\ell^*|_N$ .

The exact sequence given, together with the computation given above then yields, by [L-1 402] that $p(t) = f(t)\cdot T_k f(t)$ as claimed. □

We continue by forming $GK(F)$, the Grothendieck group associated to $K(F)$. This is the free abelian group generated multiplicatively by the irreducible polynomials in $K(F)$. $T_k$ induces an automorphism of period 2 on $GK(F)$, so we can form $H^2(C_2;GK(F))$, denoted simply $H^2(k;K(F))$ [M 122]. This is identified as $\{f \in K(F): T_k f = f\}$ modulo $\{g \in K(F): g = h\cdot T_k h\}$ . This in turn is an $F_2$-vector space with a basis element for each $T_k$ fixed irreducible polynomial. We denote this basis by $\mathcal{B}$ .

<u>Lemma 1.2</u> <u>The map</u> $X : W(k,F) \to H^2(k,K(F))$ <u>given by</u>: $[M,B,\ell] \to$ <u>characteristic polynomial of</u> $\ell$ <u>is an epimorphism</u>.

<u>Proof</u>: $X$ is well-defined by 1.1. To see that $X$ is onto it suffices to show that every $p(t) \in \mathcal{B}$ is in the image of $X$ .

Given $p(t)$, consider $V = F[t,t^{-1}]/(p(t)) = F(\theta)$. Let $B$ be given by: $B(x,y) = \mathrm{trace}_{F(\theta)/F} x\bar{y}$, where $-$ denotes the involution induced on $F(\theta)$ by $\theta \to k\theta^{-1}$.

$B$ is symmetric since $\mathrm{trace}\ x\bar{y} = \mathrm{trace}\ \bar{x}y$. $B$ is clearly non-singular; one may apply the trace lemma 2.1 to be proved or prove it directly.

Define $\ell$: $V \to V$ by $x \to \theta x$. We compute:

$$B(\ell x,\ell y) = B(\theta x,\theta y) = \mathrm{trace}(\theta\bar{\theta}x\bar{y}) = \mathrm{trace}(ky\bar{x}) = kB(x,y).$$

Hence, $[V,B,\ell] \in W(k,F)$. Since $\theta$ satisfies $p(t)$, the minimal polynomial of $\ell$ is $p(t)$. However, $p(t)$ divides the characteristic polynomial of $\ell$, and degree $p(t)$ = degree of characteristic polynomial. Hence $p(t)$ = characteristic polynomial of $\ell$. $\square$

Given a Witt equivalence class $[M,B,\ell]$, $M$ is a $D[t]$-module by identifying $t$ with $\ell$. We now wish to decompose $[M,B,\ell]$ according to the characteristic polynomial of $\ell$. We begin with:

<u>Lemma 1.3</u> <u>Let</u> $p(t)$ <u>be</u> $T_k$ <u>fixed</u>. <u>Then we can factor</u>

$$p(t) = p_1^{r_1} \ \cdots \ p_w^{r_w}\, q_1^{s_1}\, \bar{q}_1^{s_1} \ \cdots\cdots \ q_k^{s_k}\, \bar{q}_k^{s_k} \ ,$$

<u>into irreducible factors, where</u> $\bar{q}_i$ <u>denotes</u> $T_k q_i$. <u>In this decomposition, each</u> $p_i(t)$ <u>is</u> - <u>invariant</u>, ie. $T_k$ <u>fixed, and the</u> $q_i$ <u>are not</u> - <u>invariant</u>. $\square$

<u>Lemma 1.4</u> <u>Suppose</u> $(M,B,\ell)$ <u>has characteristic polynomial</u> $p(t) = p_1(t)\ p_2(t)$, <u>with</u> $p_1(t)$, $p_2(t)$ <u>relatively prime polynomials, which are both - invariant.</u> <u>Then</u> $(M,B,\ell) \simeq (M_1,B_1,\ell_1) \oplus (M_2,B_2,\ell_2)$ <u>where</u> $\ell_i$ <u>has characteristic polynomial</u> $p_i(t)$.

<u>Proof</u>: Let $M_i = \{v \in M: p_i(\ell)(v) = 0\}$, $B_i = B|_{M_i}$, $\ell_i = \ell|_{M_i}$.

Since $(p_1,p_2) = 1$, we can find polynomials $r(t)$, $s(t)$ with $p_1(t)r(t) + p_2(t)s(t) = 1$.

<u>Remark 1.5</u>: This statement is false over Z, and is the reason the decomposition fails over Z.

Hence, if $v \in M$, then $p_1(\ell)r(\ell)v + p_2(\ell)s(\ell)v = v$. However, $p_1(\ell)r(\ell)v \in M_2$ since $p_1(\ell)p_2(\ell)v = 0$. It follows that $M = M_1 + M_2$.

If $v \in M_1 \cap M_2$, by the above it is clear that $v = 0$. It follows that $M$ is a direct sum of $M_1$, $M_2$, ie. $M = M_1 \oplus M_2$.

We next show that $B = B_1 \oplus B_2$, so that $B_1$ and $B_2$ are inner products.

To begin with, consider $p_2(\ell): M_1 \to M_1$. $p_2(\ell)$ is clearly 1-1, and hence an isomorphism as we are working over a field. Thus, if $v \in M_1$, we may write $v = p_2(\ell)v_1$. Let $w \in M_2$.

$$\begin{aligned}
B(v,w) &= B(p_2(\ell)v_1,w) \\
&= B(v_1,p_2(\ell)^*w) \\
&= 1/k^n \quad B(\ell^n v_1,\ell^n p_2(\ell)^*w) \\
&= 1/k^n \quad B(\ell^n v_1,a_0 p_2(\ell)w) \\
&= 0
\end{aligned}$$

where $a_0$ = constant term of $p_2(t)$, since $p_2(t)$ is - invariant.

Thus, $B = B_1 \oplus B_2$ as claimed.

Finally, we must show $\ell_i: M_i \to M_i$. Note that $p_i(\ell)(\ell v) = \ell(p_i(\ell)(v)) = \ell(0) = 0$. Hence, $\ell_i = \ell|_{M_i}$ maps $M_i \to M_i$. This shows

$$(M,B,\ell) \cong (M_1,B_1,\ell_1) \oplus (M_2,B_2,\ell_2)$$

as claimed. □

Lemma 1.6 Suppose $(M,B,\ell)$ has characteristic polynomial $p(t)$ = characteristic polynomial of $\ell$ and $p = q_1^{r_1}\bar{q}_1^{r_1}$, where $q_1$ is irreducible. Then $(M,B,\ell) \sim 0$.

Proof: We are assuming $q_1 \neq \bar{q}_1$. Since $(q_1,\bar{q}_1) = 1$, M will split as $M = M_1 \oplus M_2$, as in 1.4. We must now examine B.

Let $v \in M_1$. As in 1.4, we may write $v = q_1^r(\ell)(v_1)$. Let $w \in M_1$.

$$\begin{aligned} B(v,w) &= B(q_1^r(\ell)v_1, q_1^r(\ell)w_1) \\ &= B(v_1, q_1^r(\ell)^* q_1^r(\ell)w_1) \\ &= 0 \end{aligned}$$

as before. Thus, B has matrix $\begin{pmatrix} 0 & - \\ - & 0 \end{pmatrix}$. Since B is non-singular, $\dim M_1 = \dim M_2 = 1/2 \dim M$. As in 1.4, $M_1$ is $\ell$ invariant. It follows that $M_1$ is a metabolizer for $(M,B,\ell)$. □

We are almost ready to state the Decomposition Theorem. First, some notation.

<u>Definition 1.7</u> <u>Let</u> T <u>be a multiplicative subset of</u> D[t]. <u>Then</u> W(k,K;T), <u>respectively</u> A(K,K;T), <u>denotes Witt equivalence classes in</u> W(k,K), A(k,K), <u>which are annihilated by</u> T. <u>In particular, when</u> T <u>consists of non-negative powers of a</u> $T_k$ <u>fixed irreducible polynomial</u> f, <u>we shall use the notation</u> W(k,K;f).

<u>Theorem 1.8</u> (<u>The Decomposition Theorem</u>) <u>For</u> F <u>a field</u> $W(k,F) \approx \bigoplus_{f \in \mathcal{B}} W(k,F;F)$ <u>where</u> $\mathcal{B}$ <u>denotes the basis of</u> $H^2(k;K(F))$ <u>consisting of</u> $T_k$ <u>fixed irreducible polynomials</u>.

<u>Proof</u>: Let $[M,B,\ell] \in W(k,F)$. Let p(t) be the characteristic polynomial of $\ell$. By 1.3, we can factor p as

$$p(t) = p_1^{r_1} \cdots p_w^{r_w}(q_1^{s_1} \bar{q}_1^{s_1}) \cdots (q_k^{s_k} \bar{q}_k^{s_k}) .$$

By induction and 1.4, $[M,B,\ell] = \oplus [M_i,B_i,\ell_i]$, where each $\ell_i$ has characteristic polynomial $p_i^{r_i}$ or $(q_i^{s_i} \bar{q}_i^{s_i})$. By 1.6, those $[M_i,B_i,\ell_i]$ with characteristic polynomial $(q_i^{s_i} \bar{q}_i^{s_i})$ are Witt ~ 0. This defines a homomorphism L: $W(k,F) \to \bigoplus_{f \in \mathcal{B}} W(k,F;f)$.

We must show L is well-defined. So suppose $[M,B,\ell]$ is metabolic. Then $(\oplus [M_i,B_i,\ell_i]) = 0$ where $\ell_i$ has characteristic polynomial $p_i^{r_i}$. We need to show that if $[M_1,B_1,\ell_1] \oplus [M_2,B_2,\ell_2] = 0$, where $p_i(t)$ is the characteristic polynomial of $\ell_i$, and $p_1(t)$ is relatively prime to $p_2(t)$, then $M_1 \sim 0$ and $M_2 \sim 0$.

We identify $M_1$ with $M_1 \oplus 0 \subset M_1 \oplus M_2$. Let H be a metabolizer for $M_1 \oplus M_2$. Then $H$ is $\ell = \ell_1 \oplus \ell_2$ invariant. Further, since $p_2(\ell) = p_2\ell_1 \oplus p_2\ell_2$, $p_2(\ell)H \subset M_1$. In fact, since $p_1(t)$ and $p_2(t)$ are relatively prime, it follows that $p_2(\ell)$ is a 1-1 mapping: $H \cap M_1 \to H \cap M_1$. Since we are working over a field, $p_2(\ell)(H \cap M_1) = H \cap M_1$. We claim $(H \cap M_1) = (H \cap M_1)^{\perp}$ in $M_1$ so that $M_1 \sim 0$ and $L$ is well-defined.

Clearly $H \cap M_1$ is $\ell_1$ invariant, and $H \cap M_1 \subset (H \cap M_1)^{\perp}$.

Let $x \in (H \cap M_1)^{\perp}$. We must show $x \in H \cap M_1$.

To begin with, note that if $(h_1,h_2) \in H$, then so is $(p_2(\ell_1)h_1,0)$ since $H$ is $\ell_1 \oplus \ell_2 = \ell$ invariant. Further, since $p_2(\ell_1)$ is an isomorphism on $H \cap M_1$, it follows that $(h_1,0) \in H$.

Now $(x,0) = x \in (H \cap M_1)^{\perp}$. If $h = (h_1,h_2) \in H$, it follows that $(h_1,0) \in H \cap M_1$. Thus

$$(B_1 \oplus B_2)((x,0),(h_1,h_2)) = B((x,0),(h_1,0)) = 0 \ .$$

Hence, $x \in H^{\perp} = H$, and $(H \cap M_1)^{\perp} \subset H \cap M_1$.

$L$ is clearly onto by 1.2.

$L$ is 1-1, since if $\oplus[M_i,B_i,\ell_i]$ has each $M_i \sim 0$ then so too is $\oplus[M_i] = 0$. □

Let us give another interpretation of this isomorphism $L$. Let $f(t)$ be a $T_k$ fixed irreducible polynomial, so $f \in \mathcal{B}$. Let $S = D[t] - (f(t))$, and $[M,B,\ell] \in W(k,K)$. Then localizing with respect to $S$, we obtain, $(M(S),B_S,\ell_S)$. Note that the adjoint map,

$$\mathrm{Ad}_R B_S\colon\ M(S) \to (\mathrm{Hom}_D(M,K))(S) = \mathrm{Hom}_{D(S)}(M(S),K(S))$$

is an isomorphism. $Ad_R B_S$ is an isomorphism since localization is an exact functor, [A,Mc 39]. The second isomorphism follows from [B-2 II 2.7].

$M$ is a torsion $D[t]$-module. Thus $M(S)$ is annihilated by $f^i(t)$, some $i$. Hence $(M(S), B_S, \ell_S) \in W(k,K;f)$.

We combine over all $f \in \mathcal{B}$, to obtain exactly the $L$ given in Theorem 1.8. Since $L$ can be viewed as arising from localizing, we shall call $L$ the localization homomorphism.

In fact, as long as we localize at all prime ideals in $D$, or $D[t]$, where $M$ is a finitely generated torsion $D$, or $D[t]$-module, we obtain such an $L$.

__Theorem 1.9__ __Let__ $K = F/D$. __By localizing at all prime ideals__ $P$ __in__ $D$, __we obtain an isomorphism:__

$$L: W(k,K) \to \bigoplus_{P \text{ prime in } D} W(k,K(P);D(P))$$

__Here__

$$K(P) = (F/D)(P) = F/D(P).$$

__Proof__: Exactly as in 1.8. □

We should like to describe these pieces $W(k,F;f)$. In order to do this, we need some further notation.

__Definition 1.10__ __Let__ $S$ __be a__ $D$-__algebra__, __finitely generated as a__ $D$-__module__. __Then__ $W(k,K;S)$ __denotes__ __Witt equivalence classes__, $[M,B,\ell]$ __in__ $W(k,K)$ __with a compatible__ $S$-__module structure__, __meaning__

<u>there exists</u> $r \in S$ <u>with</u> $rm = \ell m$ <u>for all</u> $m \in M$.

We shall be specifically interested in the case $S = F[t,t^{-1}]/(f(t))$ where $f(t) \in \mathcal{B}$. For this $S$, observe that there is an inclusion

$$j: W(k,F;S) \to W(k,F;f).$$

Structures on the left are annihilated by $f(t)$, those on the right are annihilated by some power of $f$.

<u>Proposition 1.11</u> $j$ <u>is an isomorphism</u>.

<u>Proof:</u> $j$ is clearly 1-1. Let $(M,B,\ell)$ be an anisotropic representative of a Witt equivalence class in $W(k,F;f)$. Thus if $N \subset M$, $N \neq 0$ is $\ell$ invariant, then $N \cap N^{\perp} = 0$. It follows that $M = N \oplus N^{\perp}$, and that $(M,B,\ell) = (N,B|,\ell|) \oplus (N^{\perp},B|,\ell|)$, where $B|$, $\ell|$ denote the restrictions of $B,\ell$ to $N,N^{\perp}$. This is standard linear algebra, see [H 157]. Continuing we can write

$$M = N_1 \oplus N_2 \oplus \ldots \oplus N_r$$

as a direct sum of inner product spaces $(N_i,B_i,\ell_i)$, where each $N_i$ has no non-trivial $\ell$ invariant submodules. Such $N_i$ are called <u>irreducible</u>.

Let $T_i$ = annihilator of $N_i$ in $F[t]$. We want to show $T_i$ is a maximal ideal in $F[t]$. Suppose not. Then $T_i \underset{\neq}{\subset} S_i \underset{\neq}{\subset} F[t]$, for some ideal $S_i$.

Claim: $S_iN_i \neq N_i$. For if $S_iN_i = N_i$, we recall [K-2 50]

Theorem 76: Let R be a ring, I an ideal in R, A a finitely generated R-module satisfying $IA = A$. Then $(1+y)A = 0$ for some $y \in I$.

It follows that $(1 + y)N_i = 0$ for some $y \in S_i$. Hence, $(1 + y) \in T_i \subset S_i$, so $1 + y \in S_i$. Hence $1 \in S_i$. This contradicts $S_i \subsetneqq F[t]$. Thus $S_iN_i \neq N_i$.

$S_iN_i \neq 0$, since $S_i \neq T_i$. $S_iN_i$ is t invariant, ie. $\ell$ invariant as we identify the action of t with $\ell$, because $S_i$ is an ideal.

However, we have thus constructed a non-trivial $\ell$ invariant submodule of $N_i$. This contradicts $N_i$ being irreducible. It follows that $T_i$ is indeed a maximal ideal in F[t]. Thus $T_i = (f(t))$, and j is onto. □

Remark 1.12 Proposition 1.11 has shown the two notations given in 1.7 and 1.10 to be redundant. Nonetheless we shall use both. The notation $W(k,F;f)$ is used when we wish to stress the polynomial aspect of the mapping structure. $W(k,F;F[t]/(f(t))$ is used when we wish to stress the module structure.

Proposition 1.13 Let $K = F/D$. Then the inclusion

$$W(k,K;D/P) \xrightarrow{j} W(k,K(P);D(P))$$

is an isomorphism, where $P$ is a prime ideal in D.

Proof: Same as 1.11. □

Here $W(k,K;D/P)$ denotes equivalence classes $[M,B,\ell]$ in which $M$ has a $D/P$ module structure. $W(k,K(P);D(P))$ denotes equivalence classes in which $M$ has a $D(P)$ module structure.

For $F = Q$, $D = Z$, $D/P = F_{p^n}$ a finite field. Since $M$ is a vector space over $F_{p^n}$, $B$ must take its values in the cyclic subgroup of $Q/Z$ annihilated by $p$. By the natural choice of generator for this subgroup, namely $1/p$, we have $W(k,K;D/P) \approx W(k,F_p)$.

<u>Proposition 1.14</u> $A(F)$ <u>decomposes as</u>

$$\bigoplus_{f \in B} A(F;f) = \bigoplus_{f \in B} A(F;F[t]/(f(t))$$

<u>Proof</u>: The proof is exactly like 1.11, where now $t$ acts as $s$, the symmetry operator. □

## 2. <u>The trace lemma</u>

Given $f(t) \in B$, meaning $f$ is irreducible, $T_k$ fixed, we form the field $F[t,t^{-1}]/(f(t)) = F(\theta)$. By III.1.7, $(f(t))$ is - invariant, so there is an induced involution on $F(\theta)$. This involution is non-trivial in the type 1 situation only. We aim now to identify explicitly the group $W(k,F;F[t]/(f(t)))$. We begin with:

<u>Lemma 2.1</u> (<u>The trace lemma</u>). <u>Let</u> $R$ <u>be a commutative ring with unit</u>, $A$ <u>an</u> R-<u>algebra</u>, $E$ <u>an</u> A-<u>module</u>, <u>and</u> $F$ <u>an</u> R-<u>module</u>. <u>Then there is the following correspondence</u>:

<u>Let</u> $<,> : M \times M \to E$ <u>be a non-singular bilinear form over</u> A. <u>Let</u> $t: E \to F$ <u>be an</u> R-<u>linear map, which induces an isomorphism</u>

$\hat{t}: E \to \mathrm{Hom}_R(A,F)$ , *by* $e \to t(-e) = \hat{t}(e)$.

*Then the map* $t \circ \langle , \rangle : M \times M \to F$ *is non-singular.*

*Conversely, if* $M$ *is an A-module with non-singular form* $( , ): M \times M \to F$, $( , )$ *R-linear, then there is non-singular form* $\langle , \rangle : M \times M \to E$ *with* $t \circ \langle , \rangle = ( , )$. $\langle , \rangle$ *is A-linear.*

*Further, this correspondence preserves annihilators of submodules and the metabolic property provided the R-module structure of* $M$ *lifts compatibly to* $A$.

*Proof*: *Part 1*: Given $\langle , \rangle : M \times M \to E$, and $t: E \to F$, we wish to show $( , ) = t \circ \langle , \rangle : M \times M \to F$ is non-singular.

Let $Ad_R: M \to \mathrm{Hom}_R(M,F)$ denote the adjoint of $( , )$. We want to show $Ad_R$ is an isomorphism.

*$Ad_R$ is 1-1*: Let $m \neq 0$ be in $M$. We want to show $(-,m) \neq 0$. Since $\langle , \rangle$ is non-singular, we can find $n \in M$ with $\langle n,m \rangle \neq 0$. Now $\langle n,m \rangle \in E$ and we have $\hat{t}: E \to \mathrm{Hom}_R(A,F)$. Thus, since $\hat{t}$ is an isomorphism, $t(-\langle n,m \rangle) \neq 0$. Let $a \in A$ have $t(a\langle n,m \rangle) \neq 0$. $\langle , \rangle$ is bilinear over $A$, so

$$a\langle n,m \rangle = \langle an,m \rangle$$

Hence,

$$t(a\langle n,m \rangle) = t(\langle an,m \rangle) = (an,m) \neq 0 .$$

Thus, $(-,m) \neq 0$ as claimed, and $Ad_R$ is 1-1.

*$Ad_R$ is onto*: Let $f \in \mathrm{Hom}_R(M,F)$. For each $m \in M$, define an R-linear map $A \to F$ by $a \to f(am)$. Since $\hat{t}$ is an isomorphism, this map equals $t(-f_0(m))$ for some $f_0(m) \in E$. Now $f_0$ defines

an A-linear map $f_0\colon M \to E$. By non-singularity of $<,>$ it follows that $f_0(m) = <m,n_0>$ for some $n_0 \in M$. Combining

$$f(m) = t(f_0(m)) = t(<m,n_0>) = (m,n_0),$$

so that $Ad_R$ is onto as claimed.

<u>Part 2</u>: Let $M$ be an A-module, together with a non-singular R-linear form $(,)\colon M \times M \to F$. We need to define $<,>$ with $t\circ<,> = (,)$.

Let $(-,n_0) \in Hom_R(M,F)$. As before, for each $m \in M$ we can define an R-linear map $A \to F$ by $a \to (am,n_0)$. Again $E \simeq Hom_R(A,F)$ implies $(am,n_0) = t(af_0(m))$ for some unique $f_0(m) \in E$. Now define $<m,n_0> = f_0(m)$. Then by definition $(m,n_0) = t(f_0(m)) = t(<m,n_0>)$. $f_0$ and $<,>$ are clearly A-bilinear. We now must show $<,>$ is non-singular.

Let $Ad_R$ denote the adjoint of $<,>$, $Ad_R\colon M \to Hom_A(M,E)$. <u>$Ad_R$ is 1-1</u>: Let $m \neq 0$ be in $M$. We want to show $<-,m> \neq 0$. By non-singularity of $(,)$, we can find $n \in M$ with $(n,m) \neq 0$. Hence, $<n,m> \neq 0$, else $t(<m,n>) = 0 = (m,n)$.
<u>$Ad_R$ is onto</u>: Let $f \in Hom_A(M,E)$. Then $(t\circ f) \in Hom_R(M,F)$. By the non-singularity of $(,)$ there exists $n_0 \in M$ such that $(t\circ f)(m) = (m,n_0)$, for all $m \in M$. Hence

$$(t\circ f)(am) = t(f(am)) = t(af(m)) = (am,n_0),$$

so that by definition of $<,>$ we have $<m,n_0> = f(m)$, and $Ad_R$ is onto.

The last statement of the theorem follows from the defnitions. □

We may extend Lemma 2.1 in a special case.

*Lemma 2.2* *Suppose that* $A, E$ *given in* 2.1 *have a compatible involution* $-$, *meaning* $\overline{(ae)} = \bar{a}\bar{e}$. *Suppose also that* $t\colon E \to F$ *satisfies* $t(e) = t(\bar{e})$ *for all* $e \in E$. *Then the correspondence of Lemma* 2.1 *extends to a correspondence between Hermitian forms* $(M, \langle\,,\rangle)$ *with values in* $E$, *and symmetric forms* $(M, (\,,))$ *with values in* $F$ *which have a compatible A-module structure, meaning* $(ax,y) = (x,\bar{a}y)$.

*Proof*: If $\langle\,,\rangle$ is Hermitian,

$$a\langle x,y\rangle = \langle ax,y\rangle = \langle x,\bar{a}y\rangle = \overline{\langle \bar{a}y,x\rangle} = a\overline{\langle y,x\rangle}\ .$$

Now

$$(x,y) = t(\langle x,y\rangle) = t(\overline{\langle y,x\rangle}) = t(\langle y,x\rangle) = (y,x)\ ,$$

so that $(\,,)$ is symmetric. Also

$$(ax,y) = t(\langle ax,y\rangle) = t(\langle x,\bar{a}y\rangle) = (x,\bar{a}y)\ .$$

Conversely, let $(\,,)$ be symmetric. Then

$$\begin{aligned} t(a\langle x,y\rangle) &= t(\langle ax,y\rangle) = (ax,y) = (y,ax) = (\bar{a}y,x) \\ &= t(\langle \bar{a}y,x\rangle) = t(\bar{a}\langle y,x\rangle) = t(a\overline{\langle y,x\rangle}) \end{aligned}$$

However, $E \cong \mathrm{Hom}_R(A,F)$ via $\hat{t}$, so $\langle x,y\rangle = \overline{\langle y,x\rangle}$, and $\langle\ ,\rangle$ is Hermitian. □

We recall the identification made at the beginning of Section 2, $F[t]/(f(t)) = F(\theta)$. We are now ready to compute $W(k,F;F(\theta))$, where $f(t)$ is $T_k$ fixed and irreducible.

Theorem 2.3  $W(k,F;F(\theta)) \simeq H(F(\theta))$

Proof: We apply 2.1 and 2.2, with $R = F$, and $A = E = F(\theta)$, $t: E \to F$ is the trace homomorphism.

We must check the non-singularity condition on $t$, namely that $\hat{t}: F(\theta) \to \text{Hom}_F(F(\theta),F)$ induces an isomorphism.

We must assume $F(\theta)$ is a finite, separable extension of $F$. Thus $t(- x) \neq 0$ for $x \neq 0$ [L-1 211]. It follows that $t$ is 1-1. However, $F(\theta)$ and $\text{Hom}_F(F(\theta),F)$ are vector spaces over $F$ of the same dimension. Hence $\hat{t}$ is an isomorphism, so that we may apply 2.1 and 2.2.  □

Comment: Clearly, $t(e) = t(\bar{e})$, so that 2.2 applies. In our identification, $[V,<,>] \in H(F(\theta))$ corresponds to $[V,(,),\ell]$ $W(k,F;F(\theta))$. Here $t\circ<,> = (,)$. The map $\ell$ is recovered from Hermitian as multiplication by $\theta$, $\ell v = \theta v$.

We shall reserve the term Hermitian for the case that the involution is non-trivial. Thus 2.3 is for type 1 polynomials.

For type 2 irreducible, $f(t) = t^2 - k$. We then read Theorem 2.3 as:

$$W(k,F;F[t]/((t^2 - k))) = W(F(\sqrt{k})).$$

We may thus restate the Decomposition Theorem  1.8, as

<u>Theorem 2.4</u> <u>If</u> $k \notin F^{**}$, <u>then</u>

$$W(k,F) \simeq W(F(\sqrt{k})) \quad \bigoplus_{\substack{f \in \mathcal{B} \\ f \text{ of type } 1}} H(F[t]/(f(t)))$$

<u>If</u> $k \in F^{**}$

$$W(k,F) \simeq W(F) \oplus W(F) \quad \bigoplus_{\substack{f \in \mathcal{B} \\ f \text{ of type } 1}} H(F[t]/(f(t)))$$

<u>Remarks</u>:

(1). The Hermitian terms $\oplus$ runs over all irreducible $T_k$ fixed polynomials in $K(F)$ of type 1. The same field $F[t]/(f(t)) = F(\theta) = F(\sigma) = F[t]/(g(t))$ may be repeated.

(2). If $k \in F^{**}$, the two Witt terms correspond to the two irreducible polynomials of type 3, $t + \sqrt{k}$, $t - \sqrt{k}$.

(3). If the characteristic of $F$ is 2, then $k \in F^{**}$, so

$$W(k,F) = W(F) \bigoplus_{f \in \mathcal{B}} H(F(\theta))$$

since

$$t + \sqrt{k} = t - \sqrt{k}$$

in this case. Again, the field $F(\theta)$ may be repeated.

(4). This theorem equally applies to the skew case; simply write

$$W^{\varepsilon}(k,F) \simeq W^{\varepsilon}(F(\sqrt{k})) \bigoplus_{f \in \mathcal{B}} H^{\varepsilon}(F(\theta)), \quad \varepsilon = \pm 1.$$

Our final goal of this section is to relate this discussion to the asymmetric case. Let $[M,B] \in A(F)$. Recall the symmetry operator $s\colon M \to M$ satisfying $B(x,y) = B(y,sx)$. Consequently $B(x,y) = B(sx,sy)$, so that $s$ yields a map of degree 1 on $[M,B]$.

Again, there is the involution on $F[t,t^{-1}]$ induced by $T_k$, with $k = 1$. This extends to $F(\theta) = F[t,t^{-1}]/(f(t))$, where $f$ is an irreducible $T_k$ fixed polynomial. The induced involution on $F(\theta)$ is $\theta \to \bar{\theta} = \theta^{-1}$.

Let $S = F[t]/(f(t))$, and consider $A(F;S)$. This denotes structures $[M,B]$, in which $t$ acts as $s$ the symmetry operator. Since $A(F) \cong \bigoplus A(F;S)$ by 1.14, we wish now to compute $A(F;S)$. We do this in two ways.

<u>Theorem 2.5</u> <u>Let</u> $f(t)$ <u>be an irreducible</u> $T_k$ <u>fixed polynomial of type</u> 1 $(k = 1)$. <u>Then</u>

$$A(F;f) \cong H(F(\theta)) \cong H_\theta F(\theta).$$

<u>Proof</u>: The idea in this computation is to apply the trace lemma. We may do this either using trace: $F(\theta) \to F$, or a scaled trace: $F(\theta) \to F$.

<u>Part(a)</u>: Using trace = $t$: $F(\theta) \to F$.

Let $[M, \langle,\rangle] \in H_\theta F(\theta)$ and $[M,(,)] \in A(F;f)$.

We need to show that $\langle,\rangle$ is $\theta$ Hermitian with values in $F(\theta)$ if and only if $(,)$ is asymmetric with values in $F$, satisfying $(x,y) = (y,sx)$ with $s$ being identified with $\theta$. We are of course applying the trace lemma with the map $t$ = trace $F(\theta)/F$.

Let $(,)$ satisfy $(x,y) = (y,\theta x)$. Then

$$t(a\langle x,y\rangle) = t\langle ax,y\rangle = (ax,y) = (y,\theta ax) = (\bar{\bar{\theta}}\bar{\bar{a}}y,x)$$
$$= t\langle \bar{a}\bar{\theta}y,x\rangle = t(\bar{a}\bar{\theta}\langle y,x\rangle) = t(a\theta\overline{\langle y,x\rangle}) .$$

Hence, since $\hat{t}$ is an isomorphism $\langle x,y\rangle = \overline{\langle y,x\rangle}$, so that $\langle\ ,\rangle$ is $\theta$ Hermitian.

Conversely, suppose $\langle,\rangle$ is $\theta$ Hermitian. Then

$$(x,y) = t\langle x,y\rangle = t(\theta\overline{\langle y,x\rangle}) = t(\bar{\theta}\langle y,x\rangle)$$
$$= t\langle y,\theta x\rangle = (y,\theta x)$$

as desired.

Thus $A(F;f) = H_\theta(F(\theta))$, where $F[t]/(f(t)) = F(\theta)$.

<u>Part (b)</u>: Using the scaled trace. Since $\theta\bar{\theta} = 1$, by Hilbert 90, there exist $u \in F(\theta)$ with $u\bar{u}^{-1} = \theta$. Let $t_1: F(\theta) \to F$ be given by $x \to \text{trace}(x\bar{u}^{-1})$. It is clear that $\hat{t}_1$ is an isomorphism. Again $t$ denotes trace $F(\theta)/F$.

Now suppose $[M,\langle,\rangle] \in H(F(\theta))$.

Then

$$\begin{aligned}(x,y) &= t_1\langle x,y\rangle = t(\bar{u}^{-1}\langle x,y\rangle)\\ &= t\langle x,u^{-1}y\rangle = t\overline{\langle u^{-1}y,x\rangle}\\ &= t\langle u^{-1}y,x\rangle = t\langle y,\bar{u}^{-1}x\rangle\\ &= t\langle y,\theta u^{-1}x\rangle = t_1\langle y,\theta x\rangle\\ &= (y,\theta x).\end{aligned}$$

Conversely, suppose $(x,y) = (y,\theta x)$.

Then

$$\begin{aligned}t(a\langle x,y\rangle) &= t\langle ax,y\rangle = t_1\bar{u}\langle ax,y\rangle\\ &= t_1\langle ax,uy\rangle = (ax,uy)\\ &= (uy,a\theta x) = (u\bar{\theta}y,ax)\\ &= (\bar{u}y,ax) = t_1\bar{u}\bar{a}\langle y,x\rangle\end{aligned}$$

$$= t(\bar{a}\langle y,x\rangle)$$
$$= t(a\overline{\langle y,x\rangle})$$

Again, $t$ is non-singular, so $\langle x,y\rangle = \overline{\langle y,x\rangle}$, and $\langle,\rangle$ is Hermitian. □

Note that we can choose $u = \theta/(1+\theta)$ so $\bar{u} = 1/(1+\theta)$, and $\bar{u}^{-1} = 1+\theta$. We give both identifications in this theorem since on certain occasions it is more convenient to think of Hermitian forms as giving $A(F)$. The disadvantage is that we must use a scaled trace to make this identification.

We should also give the third identification. Namely, it follows from this theorem that

$$H(F(\theta)) = H_\theta(F(\theta)) \quad \text{via:} \quad h: \langle,\rangle \to \langle,\rangle_1$$

with $h$ defined by

$$\langle x,y\rangle_1 = \langle x,u^{-1}y\rangle \quad \text{where} \quad u\bar{u}^{-1} = \theta .$$

(a) If $\langle,\rangle_1$ is $\theta$ Hermitian,

$$\begin{aligned} \langle x,y\rangle = \langle x,uy\rangle_1 &= \theta\overline{\langle uy,x\rangle_1} \\ &= \overline{\langle \bar{\theta}uy,x\rangle_1} = \overline{\langle u\theta^{-1}y,x\rangle_1} \\ &= \overline{\langle \bar{u}y,x\rangle_1} = \overline{\langle y,ux\rangle_1} \\ &= \overline{\langle y,x\rangle} \quad , \end{aligned}$$

and $\langle\ ,\ \rangle$ is Hermitian.

(b)  If  $<,>$  is Hermitian,

$$\begin{aligned}
<x,y>_1 &= <x,u^{-1}y> = \overline{<u^{-1}y,x>} \\
&= \overline{<y,\overline{u}^{-1}x>} = \overline{<y,\theta u^{-1}x>} \\
&= \overline{\theta}\,\overline{<y,u^{-1}x>} = \theta\overline{<y,u^{-1}x>} \\
&= \theta\overline{<y,x>}_1
\end{aligned}$$

so that  $<,>_1$  is  $\theta$  Hermitian.

__Corollary 2.6__  __There is a commutative triangle of__

$$
\begin{array}{ccc}
& h & \\
H(F(\theta)) & \to & H_\theta F(\theta) \\
t_1 \searrow & & \swarrow\; t \\
& A(F;F[t]/(f(t))) &
\end{array}
$$

__isomorphisms__. t __is__ __trace__ $F(\theta)/F$. $t_1$ __is__ __trace__ __scaled__ __by__ $\overline{u}^{-1}$,
__where__  $u/\overline{u} = \theta$ . h: $<,> \to <,>_1$, __is__ __defined__ __by__ $<x,y>_1 = <x,u^{-1}y>$ . $\square$

Thus, the decomposition theorem reads,

__Theorem 2.7__  $A(F) \simeq \bigoplus_{f\in B} H_\theta(F(\theta)) \simeq \bigoplus_{f\in B} H(F(\theta))$ .

For  $F = Q$  the rationals, or  $F = F_p$  a finite field,  Hermitian
of  $F(\theta)$  is well known  [Lh]  and  [M,H] , see Chapter II 5.4 .

## 3. Computing Witt groups

We are interested in the group $W(k,Z)$. Let $[M,B,\ell] \in W(k,Z)$. Then we may view $M$ as a $Z[t,t^{-1}]/(f(t))$ module, where $f(t)$ is the characteristic polynomial of $\ell$, and $t$ acts as $\ell$. As has been pointed out, for $S = Z[t,t^{-1}]/(f(t))$, the decomposition theorem fails; $W(k,Z) \neq \oplus W(k,Z;S)$

Later, we shall measure this failure, [VIII 1]. Our next task is to describe these pieces $W(k,Z;S)$, for $S = Z[\theta]$ above. Thus, let $f$ be a monic, integral $T_k$ fixed irreducible polynomial, $S = Z(\theta) = Z[t,t^{-1}]/(f(t))$. We begin by describing the maximal ideals in $S$.

<u>Proposition 3.1</u> <u>The maximal ideals of</u> $S$ <u>are of the form</u> $M = (p,g(\theta))$, <u>where</u> $g$ <u>is a monic integral polynomial whose</u> mod p <u>reduction</u> $\gamma$ <u>is irreducible and</u> $\gamma$ <u>divides the</u> mod p <u>reduction of</u> $f$, <u>denoted</u> $\phi$.

<u>Proof</u>: $S/M$ is clearly a finite field, indeed it embeds into $D/P$ where $D$ is the maximal order, and $P \cap S = M$.

Suppose $S/M$ lies over the prime field $F_p$. It follows that $p \in M$. Further, $S/M$ is generated by $\theta_1$, the image of $\theta$ in residue field.

Let $\gamma(t)$ be the monic irreducible polynomial over $F_p$ of $\theta_1$. Let $g(t)$ be a monic integral polynomial whose mod p reduction is $\gamma(t)$. Then clearly $g(\theta) \in M$ and $(p,g(\theta)) = M$.

Since $g$ is irreducible mod p, $g$ is irreducible. Further, $f(\theta) = 0$, so $\phi(\theta_1) = 0$, and $\gamma$ divides $\phi$ as claimed. □

Remark: $M$ is invariant under the involution $-$ induced by $\bar{\theta} = k\theta^{-1}$ if and only if $S/M$ has a well-defined involution induced by $-$. Further, $S/M$ has involution $-$ if and only if $\gamma(t)$, the irreducible polynomial of $\theta_1$ is $T_k$ fixed. For if $\gamma$ is $T_k$ fixed, we have already seen there is an involution induced on $F_p(\theta_1) \simeq S/M$. Conversely, when there is the $-$ involution on $F_p(\theta_1)$, $\gamma(\theta_1) = \gamma(\bar{\theta}_1) = 0$. Equating coefficients it follows easily that $\gamma$ is $T_k$ fixed.

We wish to apply the trace lemma to compute $W(k,Z;S)$. Thus we consider the inverse different of $S$,

$$\Delta^{-1}(S/Z) = I = \{x \in E\colon \operatorname{trace}_{E/Q}(xS) \subset Z\}$$

Here $E$ is the quotient field of $S$.

Again, there is the $-$ involution on $E$, and $S$ is a $-$ invariant order. It follows that the inverse different $I$ is a $-$ invariant fractional ideal over $S$.

We may describe $I$ by Euler's theorem [A 92], namely $I = Z(\theta)/(f'(\theta))$, where $f'$ is the derivative of $f$.

We wish to apply the trace lemma with: $A = S$, $E = \Delta^{-1}(S/Z)$, $F = Z$, $t\colon E \to F$, $t = \operatorname{trace} E/F$, $R = Z$. In order to do this, we must verify that

$$t\colon \Delta^{-1}(S/Z) \to \operatorname{Hom}_Z(S,Z)$$

is an isomorphism. To begin with, the map $x \to t(-x)$ is 1-1 since it is 1-1 on the quotient fields. Continuing, let $h \in \operatorname{Hom}_Z(S,Z)$. Then $h = t(-x_0)$, for $x_0 \in E$, since $t$ is an isomorphism on the field level. However $h|_S$ is Z-valued, so that $\mathrm{trace}(x_0 S) \subset Z$, and $x_0 \varepsilon I$. Hence $t$ is onto.

Thus, there is an isomorphism between I-valued Hermitian forms, and Z-valued symmetric forms with a compatible S-module structure, meaning $(rx,y) = (x,\bar{r}y)$ for all $r \varepsilon S$. We state this as:

__Theorem 3.2__ __The trace lemma yields an isomorphism__

$$t_*: H(\Delta^{-1}(S/Z)) \to W(k,Z;S) \quad \square$$

The same result naturally holds for asymmetric, and we have:

__Theorem 3.3__ __The trace lemma yields an isomorphism__

$$H_\theta(\Delta^{-1}(S/Z)) \simeq A(Z;S). \quad \square$$

Caution: Using a scaled trace may be impossible since if $u\bar{u}^{-1} = \theta$, $u \varepsilon E$, and $u$ may not even be in $S$.

We next apply the trace lemma to the trace map $t_*: E/\Delta^{-1}(S/Z) \to Q/Z$, in other words $t_*$ is induced from trace E/Q. There is the induced isomorphism:

$$\hat{t}_*: E/\Delta^{-1}(S/Z) \to \mathrm{Hom}_Z(E/\Delta^{-1}(S/Z), Q/Z).$$

Again, the trace lemma yields, (see Remark 3.9):

__Theorem 3.4__ $H(E/\Delta^{-1}(S/Z)) \simeq W(k,Q/Z;S)$

$$H_\theta(E/\Delta^{-1}(S/Z)) \simeq A(Q/Z;S). \quad \square$$

While we are discussing $W(k,Q/Z;S)$, we continue with the analog of the Decomposition Theorem 1.9.

<u>Theorem 3.5</u> 
$$W(k,Q/Z;S) \simeq \bigoplus_{M=\bar{M}} W(k,Q/Z;S/M) \simeq \bigoplus_{M=\bar{M}} W(k,F_p;S/M) \simeq \bigoplus_{M=\bar{M}} H(S/M)$$
<u>where</u> $\oplus$ <u>runs over all - invariant maximal ideals in</u> S.

<u>Proof</u>: The proof is exactly as before. Let $[M,B,t] \in W(k,Q/Z;S)$. We write this as $\oplus [M_i,B_i,t_i]$, of irreducible modules. Thus $A_i$ = S - annihilator of $M_i = \{r \in S: rm = 0 \text{ for all } m \in M_i\}$, is a maximal ideal in S. Now we check that $A_i$ is - invariant. Observe $0 = B_i(ax,y) = B_i(x,\bar{a}y)$ for all $a \in A$, $x,y \in M_i$. Hence $\bar{a} \in A_i$, for if $\bar{a} \notin A$; $\bar{a}y \neq 0$ for some $y \in M_i$, and $0 = B(x,\bar{a}y)$ for all $x$ contradicts the non-singularity of $B_i$. The rest is as before to give the first isomorphism.

Now $S/M$ is a finite field, with induced involution since $M = \bar{M}$. Of course, this may be the trivial involution. Any finitely generated $S/M$ - module is a finite dimensional vector space whose underlying abelian group is p torsion, where $p = \text{char } S/M$. The second isomorphism then follows by selecting a generator, say $1/p$ for the p torsion in $Q/Z$. The last isomorphism follows by 2.1. □

<u>Remark 3.6</u> A similar theorem holds for $A(k,Q/Z;S)$.

<u>Remark 3.7</u> If the involution on $S/M$ is trivial, the last

term is actually $W(S/M)$, Witt of a finite field, which is determined by the cardinality of $S/M$. Let $q$ = cardinality of $S/M$.

(a) If $q \equiv 1 \pmod 4$ $W(S/M) \simeq Z/2Z \oplus Z/2Z$

(b) If $q \equiv 0 \pmod 2$ $W(S/M) \simeq Z/2Z$

(c) If $q \equiv 3 \pmod 4$ $W(S/M) \simeq Z/4Z$

If the involution on $S/M$ is non-trivial, we have Hermitian of a finite field. Here rank is the only invariant [M-H 117].

Remark 3.8 We have thus shown that $H(E/I) \simeq \bigoplus_{M=\bar{M}} H(S/M)$.

In fact, this holds directly, when $I = \bar{I}$ is a - invariant fractional ideal with $S = D$ the underlying ring of integers in $E$. For if $[M,B] \in H(E/I)$, with $M$ a finitely generated torsion $D$-module, we take anisotropic representatives, decompose into irreducibles, etc. It follows that $[M,B] = \oplus [M_i,B_i]$, where the annihilator of $M_i$, say $P$, is a - invariant maximal ideal in $D$. Thus $B_i$ takes values in $E/I(P)$. We may identify this with a $D/P$-valued form [VI 3]. For $I = \Delta^{-1}(D/Z)$, this was done in 3.5. In general, we embed $D/P$ into $E/I(P)$ by $r + P \to \rho r + I(P)$, where $\rho$ has valuation $v_P(I) - 1$, [VI 3], and obtain an isomorphism between Hermitian forms $[M,B]$ with values in $D/P$, and Hermitian forms $[M,B]$, where $M$ is $D/P$-module, with values in $E/I(P)$.

Remark 3.9 The Hermitian groups $H(E/\Delta^{-1}(S/Z))$ are defined because $E/\Delta^{-1}(S/Z)$ is an injective $S$-module, for $S$ an order. One verifies this using the trace induced map from $E/\Delta^{-1}$ to $Q/Z$.

## 4. Torsion in W(-k,F)

While we are computing Witt groups, it seems natural to mention the torsion in W(-k,F). Rather surprisingly this is related to the number of squares needed to express k as a sum of squares.

**Theorem 4.1** *If k is a square in F, then W(-k,F) is all 2-torsion.*

*Proof*: Suppose $k = r^2$, and let $[M,B,\ell] \in W(-k,F)$. Then clearly $N = \{(rx,\ell x): x \in M\}$ is a metabolizer for $[M,B,\ell] \oplus [M,B,\ell] = 2[M,B,\ell]$. □

**Theorem 4.2** *If k is a sum of two squares in F, then W(-k,F) is all 4-torsion.*

*Proof*: Suppose $k = r^2 + s^2$, and let $[M,B,\ell] \in W(-k,F)$. We want to show $4[M,B,\ell] = 0$. So consider $N \subset M \oplus M \oplus M \oplus M$ defined by: N = subspace generated by $\{(rx,sx,\ell x,0), (-sy,ry,0,\ell y): x,y \in M\}$. It is easy to see that N is a metabolizer, so $4[M,B,\ell] = 0$. □

**Theorem 4.3** *If k is a sum of four squares in F, then W(-k,F) is all 8-torsion.*

*Proof*: Suppose $k = r^2 + s^2 + t^2 + u^2$, and let $[M,B,\ell]$ W(-k,F). As above we produce a metabolizer for $8[M,B,\ell]$.

Let $N$ = subspace generated by

$$\{(rx,sx,tx,ux,\ell x,0,0,0),\ (-sy,ry,-uy,ty,0,\ell y,0,0),\ (-tz,uz,rz,-sz,0,0,\ell z,0),\ (-uw,-tw,sw,rw,0,0,0,\ell w):\ x,y,z,w \in M\}.$$

Cearly $N = N^{\perp}$ is a metabolizer for $8[M,B,\ell]$. □

As long as $k$ is an integer this completes the discussion since every positive integer is a sum of at most 4 squares. A few questions remain however:

(1) If $k$ is a sum of 4-squares for example, could $W(-k,F)$ in fact be all 2-torsion? How would one recognize this?

(2) In general we have only demanded $k \in F$. Thus there remains open the question of whether one can relate the torsion in $W(-k,F)$ to the minimum number of squares needed to express $k$ as a sum, for arbitrary $k$.

The technique used to prove Theorems 4.1-4.3 is basically to construct an orthogonal design of type (1,1,1,...) on the independent variables $r,s,t,\ldots$ see [G-S], and use this to construct a metabolizer N. We can thus extend Theorem 4.3 to the case of $k$ being a sum of 8 squares by constructing an orthogonal design using the Cayley numbers. However, by a Theorem of Radon this is absolutely as far as this method will go. The reader is referred to [G-S] for a discussion of orthogonal designs. However, the fact that our method of constructing metabolizers does not generalize still yields no information as to whether Theorems 4.1-4.3 generalize or not.

Chapter V THE SQUARING MAP

We wish now to study the squaring map $S_\varepsilon : W^\varepsilon(k,K) \to W^\varepsilon(k^2,K)$, where $K$ is a field, or Dedekind domain. $S_\varepsilon$ is defined by:

$$[M,B,\ell] \to [M,B,\ell^2].$$

Here $\varepsilon = +1$ if $B$ is symmetric; $\varepsilon = -1$ if $B$ is skew-symmetric. To begin with we study $S_\varepsilon$ for $K = F$ a field. We shall relate this to the case of $K = Z$ the integers in the ensuing chapters.

We shall derive an exact sequence involving the groups $A(F)$, and $W^\varepsilon(-k,F)$. The octagon we obtain is:

$$\begin{array}{ccccccccc}
 & & W^1(k,F) & \xrightarrow{S_1} & W^1(k^2,F) & \xrightarrow{I_1} & W^1(-k,F) & & \\
 & {\scriptstyle m_1}\nearrow & & & & & & \searrow{\scriptstyle d_1} & \\
A(F) & & & & & & & & A(F) \\
 & {\scriptstyle d_{-1}}\nwarrow & & & & & & \swarrow{\scriptstyle m_{-1}} & \\
 & & W^{-1}(-k,F) & \xleftarrow{I_{-1}} & W^{-1}(k^2,F) & \xleftarrow{S_{-1}} & W^{-1}(k,F) & &
\end{array}$$

$m_\varepsilon : A(F) \to W^\varepsilon(k,F)$ is defined by:

$[M,B] \to [M \oplus M, B_\varepsilon, \phi_\varepsilon]$ where

$B_\varepsilon((x,y),(z,w)) = B(x,w) + \varepsilon B(z,y)$ and

$\phi_\varepsilon(x,y) = (\varepsilon k s^{-1} y, x)$

$I_\varepsilon$: $W^\varepsilon(k^2,F) \rightarrow W^\varepsilon(-k,F)$ is defined by:

$[M,B,\ell] \rightarrow [M \oplus M,\ B \oplus -kB, \tilde{\ell}]$ where $\tilde{\ell}(x,y) = (\ell y, x)$.

$d_\varepsilon$: $W^\varepsilon(-k,F) \rightarrow A(F)$ is defined by:

$[M,B,\tilde{\ell}] \rightarrow [M,\bar{B}]$ where $\bar{B}(x,y) = k^{-1}B(x,\ell y)$.

This octagon is only defined for $k \neq 0$ in F. When $k = 0$, the maps do not all make sense.

Remark 0.1 If $k = 0$, $W(k,F) = W(F)$. To see this, let $[M,B,\ell] \in W(k,F)$ and suppose $(M,B,\ell)$ is anisotropic. We claim $\ell = 0$.

Consider the $\ell$ invariant subspace of M generated by: $\{\ell M, \ell^2 M, \ldots, \ell^t M\}$, where t = degree of char polynomial of $\ell$. This subspace is self annihilating since $B(\ell x, \ell y) = kB(x,y) = 0$. It follows that it must be 0 since $(M,B,\ell)$ is anisotropic. Hence $\ell M = 0$ as claimed.

Remark 0.2 For $k = \pm 1$, the maps given are well-defined over Z. The key observation is that the image spaces actually are inner products.

For the map $m_\varepsilon$, let $[W,B] \in A(Z)$, with symmetry operator s. Using Theorem I 4.10 we see that $s^* = s^{-1}$, so that $s^{-1}$ exists and s is non-singular. Thus $m_\varepsilon$ makes sense over Z, and $\phi_\varepsilon$ is well-defined.

$S_\varepsilon$ and $I_\varepsilon$ are clearly well-defined.

Let $[M,B,\ell] \in W^\varepsilon(-k,Z)$. As above, $\ell$ is non-singular and $\bar{B}(x,y) = k^{-1}B(x,\ell y)$; so $\bar{B}$ is non-singular when $k = \pm 1$.

In Section 2, we shall prove exactness of this octagon over $F$ a field. The proof given does not work over $Z$. We shall develop the machinery to study this problem in the following chapters.

In Section 1 we motivate the methods used in Section 2 by deriving the transfer sequence of Scharlau, Elman and Lam. In fact, the exact octagon we obtain is a generalization of these maps. The Scharlau, Elman, Lam transfer sequence is an exact octagon with several terms vanishing. We shall prove the sequence below is exact:

$$\begin{array}{ccccccc}
 & W^{+1}(F(\sqrt{a})) & \overset{S}{\rightarrow} & W^{+1}(F) & \overset{I}{\rightarrow} & W^{+1}(-a,F;f) & \\
m \nearrow & & & & & & \searrow \\
W^{+1}(F) & & & & & & 0 \\
d \nwarrow & & & & & & \swarrow \\
 & W^{-1}(-a,F;f) & \leftarrow & 0 & \leftarrow & 0 &
\end{array}$$

where $f(t) = t^2 - a$. The term $W^{\varepsilon}(-a,F;f)$ with $f(t) = t^2 - a$ denotes Witt equivalence classes of triples $(M,B,\ell)$ with:

(a) $B$ $\varepsilon$ symmetric $B: M \times M \rightarrow F$

(b) $B(\ell x,\ell y) = -aB(x,y)$

(c) $\ell$ satisfies $\ell^2 = a$. It follows that $\ell^* = -\ell$. This is consistent with our previous notation. By the trace lemma, $W^{\varepsilon}(-a,F;f) = H^{\varepsilon}(F(\sqrt{a}))$.

The maps in this sequence are given by:

$d: W^{-1}(-a,F;f) \rightarrow W(F)$ $\quad [M,B,\ell] \rightarrow [M,\overline{B}]$ where $\overline{B}(x,y) = B(x,\ell^{-1}y)$

$m: W(F) \rightarrow W(F(\sqrt{a}))$ $\quad [M,B] \rightarrow [M,B] \otimes_F F(\sqrt{a})$

$S: W(F(\sqrt{a})) \to W(F)$ is Scharlau's transfer $[M,B] \to [M,\overline{B}]$

where $\overline{B}(x,y) = tB(x,y)$

$t: F(\sqrt{a}) \to F$ is the scaled trace

[Lm 201] $t(x) = \text{trace}\,(\frac{\sqrt{a}}{2a}x)$

$I: W(F) \to W^{+1}(-a,F;f)$ $[M,B] \to [M \oplus M, \tilde{B}, \tilde{\ell}]$

$\tilde{B}((x,y),(u,v)) = B(x,u) - aB(y,v)$

$= B \oplus -aB((x,y),(u,v))$

$\tilde{\ell}(x,y) = (ay,x)$

We shall begin with the case above, and discuss the Scharlau, Elman, Lam transfer sequence.

1. Scharlau's transfer

Let $F$ be a field, $a \notin F^{**}$. Then we can form $F(\sqrt{a})$. Consider $W(F)$. This is Witt equivalence classes of pairs $[M,B]$, where $B: M \times M \to F$ is a symmetric inner product.

There is the map $m: W(F) \to W(F\sqrt{a})$ given by $[M,B] \to [M,B] \otimes_F F(\sqrt{a})$. Likewise, there is a map $S: W(F(\sqrt{a}) \to W(F)$ given by $[M,B] \to [M,\overline{B}]$ where $\overline{B}(x,y) = t \circ B(x,y)$ with $t$ defined by $t(1) = 0$, $t(\sqrt{a}) = 1$. This map $S$ is called Scharlau's transfer. It arises from a scaled trace, namely $t(x) = \text{trace}_{F(\sqrt{a})/F}(\frac{\sqrt{a}}{2a}\cdot x)$.

In [Lm ]01] an exact sequence involving $m$ and $S$ is discussed. $W(F) \overset{m}{\to} W(F(\sqrt{a}) \overset{S}{\to} W(F)$. We examine this sequence in a setting which will generalize to our situation. In so doing, we compute the cokernel of $S$.

To begin with, let us identify $M \otimes_F F(\sqrt{a})$ with $M \oplus M$ as a vector space over $F(\sqrt{a})$. Thus, if $\{v_i\}$ is a basis for $M$ over

$F$, $\{(v_i,0)\}$ will be a basis for $M \oplus M$ over $F(\sqrt{a})$. We must be careful about scalar multiplication.

We are viewing $\sqrt{a}$ as $(0,1)$, so that

$$\begin{aligned}
(1,0) &\cdot (v_i,0) = (v_i,0)\\
(0,1) &\cdot (v_i,0) = (0,v_i)\\
(1,0) &\cdot (0,v_i) = (0,v_i)\\
(0,1) &\cdot (0,v_i) = (av_i,0)
\end{aligned}$$

In other words, scalars from $F(\sqrt{a})$ are ordered pairs $(c,d)$, $c,d \in F$ to be identified with $c + d\sqrt{a}$. They operate on $M \oplus M$ as described above.

Under this identification, we can view the map $m$ as defined by:

$$m\colon [M,B] \to [M \oplus M,B'] \quad \text{where}$$
$$B'((x_1,y_1),(x_2,y_2)) = B(x_1,x_2) + aB(y_1,y_2) + [B(x_1,y_2) + B(x_2,y_1)]\sqrt{a}.$$

With these preliminaries, we proceed to define the maps involved in the transfer exact sequence.

Let $f(t) = t^2 - a$, and form $W^{-1}(-a,F;f)$. The maps are defined by:

$$\begin{aligned}
d\colon W^{-1}(-a,F;f) &\to W(F) & [M,B,\ell] &\to [M,\bar{B}] \quad \bar{B}(x,y) = B(x,\ell^{-1}y)\\
m\colon W(F) &\to W(F\sqrt{a})) & [M,B] &\to [M,B] \otimes_F F(\sqrt{a})\\
S\colon W(F(\sqrt{a})) &\to W(F) & [M,B] &\to [M,\bar{B}]
\end{aligned}$$

$$\bar{B}(x,y) = t \circ B(x,y)$$

$$t = \text{scaled trace}$$

$$I: W(F) \to W^{+1}(-a,F;f) \qquad [M,B] \to [M+M,\tilde{B},\tilde{\ell}]$$
$$\tilde{B} = B \oplus -aB$$
$$\tilde{\ell}(x,y) = (ay,x)$$

We begin by studying the map $d$.

$$d: [M,B,\ell] \to [M,\bar{B}] \qquad \bar{B}(x,y) = B(x,\ell^{-1}y)$$

$\tilde{B}$ is symmetric since
$$\begin{aligned} B(x,y) &= B(x,\ell^{-1}y) \\ &= -B(x,-\ell^{-1}y) \\ &= B(-\ell^{-1}y,x) && \text{since } B \text{ is skew-symmetric} \\ &= B(y,\ell^{-1}x) && \text{since } \ell^* = -\ell \\ &= \tilde{B}(y,x). \end{aligned}$$

Lemma 1.1 $\ker d = 0$.

Define $I: W^{-1}(F) \to W^{-1}(-a,F;f)$ by $[M,B] \to [M \oplus M,\tilde{B},\tilde{\ell}]$ where $\tilde{B}((x,y),(u,v)) = B(x,u) - aB(y,v)$ $\tilde{\ell}(x,y) = (ay,x)$. $I$ is clearly a well-defined group homomorphism. We shall show $\ker d \subsetneq \operatorname{im} I$. However, $W^{-1}(F) = 0$, so this will show that kernel $d = 0$.

Let $(M,B,\ell)$ be an anisotropic representative of a Witt class in $W^{-1}(-a,F;f)$, with $d([M,B,\ell]) = 0$. Let $N$ be a metabolizer for $[M,\bar{B}]$.

Consider $[N,B_1]$, where $B_1 = B|_N$, the restriction of $B$ to $N$. $B_1$ is non-singular since $(M,B,\ell)$ is anisotropic over a field, and $N$ is a metabolizer for $\bar{B}$. $B_1$ is a skew-symmetric inner product on $N$. Applying $I$ we obtain $[N \oplus N,\tilde{B}_1,\tilde{\ell}]$, where $\tilde{B}_1(x,y) = B_1 \oplus -aB_1$.

Define $\gamma: N \oplus N \to M$ by $(n_1,n_2) \to n_1 + \ell n_2$. We shall show $\gamma$ is an equivariant isomorphism, hence $[M,B,\ell] \in \text{im } I$, which completes the proof.

Since $\dim(N \oplus N) = \dim M$, in order to show that $\gamma$ is an isomorphism, it suffices to show $\gamma$ is $1-1$, since these are vector spaces. So suppose $\gamma(n_1,n_2) = n_1 + \ell n_2 = 0$. Then form $W = \langle n_2, \ell n_2 \rangle$, the subspace generated by $n_2$ and $\ell n_2$. $W$ is $\ell$ invariant since $\ell^2 = a$.

We compute:

$$\begin{aligned} B(n_2,n_2) &= B(n_2,-\ell^{-1}n_1) \\ &= \bar{B}(n_2,n_1) = 0 \quad \text{since } N = N^{\perp} \text{ with respect to } \bar{B} \\ B(n_2,\ell n_2) &= B(-\ell^{-1}n_1,-n_1) \\ &= -B(n_1,\ell^{-1}n_1) = 0 \end{aligned}$$

Similarly, $B(\ell n_2,\ell n_2) = 0$, and we see that $W$ is an $\ell$ invariant subspace of $M$ with $W \subset W^{\perp}$. This contradicts $(M,B,\ell)$ being anisotropic unless $W = 0$. Thus $n_2 = 0$ and since $n_1 + \ell n_2 = 0$, $n_1 = 0$. Hence $\gamma$ is an isomorphism.

The following computations show that $\gamma$ is equivariant:

$$\begin{aligned} \tilde{B}_1((n_1,n_2),(n_1',n_2')) &= B_1(n_1,n_1') - aB_1(n_2,n_2') \\ &= B(n_1,n_1') + B(\ell n_2,\ell n_2') + B(n_1,\ell n_2') + B(\ell n_2,n_1') \\ &\quad (\text{Since } N = N^{\perp} \text{ with respect to } \bar{B}) \\ &= B(n_1 + \ell n_2, n_1' + \ell n_2') \\ &= B(\gamma(n_1,n_2),\gamma(n_1',n_2')) \end{aligned}$$

$$\gamma \circ \tilde{\ell}(n_1,n_2) = \gamma(an_2,n_1) = an_2 + \ell n_1$$

$$\ell \circ \gamma(n_1,n_2) = \ell(n_1 + \ell n_2) = \ell n_1 + \ell^2 n_2 = \ell n_1 + an_2.$$

Thus $\gamma$ yields an equivariant isomorphism $(N \oplus N, \tilde{B}_1, \tilde{\ell}) \to (M,B,\ell)$. It follows that $[M,B,\ell] \in \text{im } I$ as was shown.

The natural second step is to continue computing kernels. We could do this formally although it would amount to computing the kernel of the 0 mapping:

$$I: W^{-1}(F) \to W^{-1}(-a,F;f)$$

where $W^{-1}(F) = 0$.

We shall return to I later, as this "same map" occurs at the end of our octagon.

Thus, we next study the cokernel of d, and the map

$$m: W(F) \to W(F(\sqrt{a})).$$

As observed previously we may view m as being defined by

$$[M,B] \to [M \oplus M, B']$$

where
$$B'((x_1,y_1),(x_2,y_2)) = B(x_1,x_2) + aB(y_1,y_2) + [B(x_1,y_2) + B(x_2,y_1)]\sqrt{a}.$$

<u>Lemma 1.2</u> ker m = im d.

<u>Step 1</u>: im d $\subseteq$ ker m. Suppose $[M,B,\ell] \xrightarrow{d} [M,\bar{B}] \xrightarrow{m} [M \oplus M,\bar{B}']$. Consider $N = \{(\ell v,v): v \in M\} \subset M \oplus M$

$$\begin{aligned}\overline{B}'((\ell v,v),\ (\ell w,w)) &= \overline{B}(\ell v,\ell w) + a\overline{B}(v,w) + \\ &\quad [\overline{B}(\ell v,w) + \overline{B}(\ell w,v)]\sqrt{a} \\ &= B(\ell v,w) + aB(v,\ell^{-1}w) + \\ &\quad [B(\ell v,\ell^{-1}w) + B(\ell w,\ell^{-1}v)]\sqrt{a} \\ &= -aB(v,\ell^{-1}w) + aB(v,\ell^{-1}w) + \\ &\quad [B(v,-w) + B(-w,v)]\sqrt{a} \\ &= 0\end{aligned}$$

Thus $N \subset N^{\perp}$, rank N = rank M, and rank N + rank $N^{\perp}$ = rank $M \oplus M$. Hence rank $N^{\perp}$ = rank M = rank N, and $N = N^{\perp}$, so $[M \oplus M, \overline{B}'] = 0$.

<u>Step 2.</u> ker m $\subseteq$ im d. Let (M,B) be anisotropic. Suppose $m[M,B] = 0$. Let N be a metabolizer for $[M \oplus M, B']$. Consider $K = \{x \in M: (x,0) \in N\}$. $B'((x,0),\ (y,0)) = B(x,y) = 0$ for all $x,y \in N$. Thus $K \subset K^{\perp}$, so that $K = 0$ since (M,B) is anisotropic.

Similarly, $\{x \in M: (0,x) \in N\} = 0$, and we can conclude that N is the graph of a 1 - 1 function

$$\ell: M \rightarrow M, \quad \text{i.e. } N = \{(\ell x,x): x \in M\}.$$

$$B'((\ell x,x),\ (\ell y,y)) = B(\ell x,\ell y) + aB(x,y) + [B(\ell x,y) + B(\ell y,x)]\sqrt{a} = 0.$$

Hence $B(\ell x,\ell y) = -aB(x,y)$. $B(\ell x,y) = -B(\ell y,x) = B(x,-\ell y)$. Thus $\ell$ is of degree $-a$, and $\ell^* = -\ell$. Further,

$$-aB(x,y) = B(\ell x,\ell y) = -B(x,\ell^2 y)$$

so that $\ell^2 = a$.

Now we form the space $(M,B_1,\ell)$, where $B_1$ is defined by $B_1(x,y) = B(x,\ell y)$. In order to show $B_1$ is non-singular, we claim: $B_1(-,x) = 0$ implies $x = 0$. Proof: $B_1(-,x) = 0 = B(-,\ell x)$, hence $\ell x = 0$ since $B$ is non-singular. Thus $x = 0$ since $\ell$ is $1-1$. Since we are over a field, $Ad_R B_1$ $1-1$ implies $Ad_R B_1$ is an isomorphism, and $B_1$ is non-singular.

$B_1$ is skew-symmetric since:

$B_1(x,y) = B(x,\ell y) = -B(\ell x,y) = -B(y,\ell x) = -B_1(y,x)$.

$\ell$ is of degree $-a$ with respect to $B_1$ since:

$B_1(\ell x,\ell y) = B(\ell x,\ell^2 y) = -aB(x,\ell y) = -aB_1(x,y)$.

Thus $[M,B_1,\ell] \in W^{-1}(-a,F;f)$, where $f(t) = t^2-a$.
Applying $d$ we obtain $[M,\bar{B}_1]$. $\bar{B}_1(x,y) = B_1(x,\ell^{-1}y) = B(x,y)$.
Hence $[M,B] \in \text{im } d$ as required. □

Lemma 1.3 ker $S$ = im $m$.

Step 1: im $m \subseteq$ ker $S$.
Let $m([M,B]) = [M \oplus M, B']$. Applying $S$ we obtain $[M \oplus M, t \circ B']$. Now consider $M \oplus 0 \subset M \oplus M$. By definition of $t \circ B'$, we have $M \oplus 0 \subseteq (M \oplus 0)^{\perp}$. However, $\dim (M \oplus 0) = \frac{1}{2} \dim (M \oplus M)$, so that $M \oplus 0$ is a metabolizer, and $[M \oplus M, t \circ B'] = 0$.

Step 2: ker $S \subseteq$ im $m$.

Let $(M,B)$ be anisotropic.

Suppose $S[M,B] = [M,t \circ B] = 0$. Recall $t$ is the scaled trace of $F(\sqrt{a})$ over $F$, trace $(\frac{\sqrt{a}}{2a} -)$. Let $N$ be a metabolizer for $[M,t \circ B]$.

If $c + d\sqrt{a} \in F(\sqrt{a})$, where $c, d \in F$, we shall call $c$ the <u>F-part</u>, $d$ the <u>$\sqrt{a}$-part</u> of $c + d\sqrt{a}$. $t: F(\sqrt{a}) \to F$ is given by $c + d\sqrt{a} \overset{t}{\to} d$, projection to the $\sqrt{a}$-part.

Consider $[N, B|_N] \in W(F)$. Applying $m$, we obtain $[N \oplus N, B']$. Define $\gamma : N \oplus N \to M$ by $(n_1, n_2) \to n_1 + \sqrt{a}\, n_2$. We shall show that $\gamma$ is an isomorphism of $(N \oplus N, B')$ with $(M,B)$, and hence $\ker S \subseteq \operatorname{im} m$.

<u>Comment</u>: In order that $[N, B|_N] \in W(F)$, we should again check that $B|_N$ is non-singular. This follows as in 1.1 since $(M,B)$ is anisotropic.

An alternate proof can be given since $\gamma$ is an equivariant isomorphism. Since $B$ is non-singular, so is $B'$, and hence so is $B|_N$.

In order to show $\gamma$ is an equivariant isomorphism, we first show that $\gamma$ is $1 - 1$, and hence an isomorphism as we are working over a field.

Suppose $\gamma(n_1, n_2) = n_1 + \sqrt{a}n_2 = 0$. Consider $\langle n_1 \rangle \subseteq M$. $B(n_1, n_1)$ has $\sqrt{a}$-part 0 since $N = N^{\perp}$ with respect to $\bar{B} = t \circ B$. However, $B(n_1, n_1) = B(n_1, -\sqrt{a}n_2) = -\sqrt{a}B(n_1, n_2)$. Observe that the $\sqrt{a}$-part of $B(n_1, n_2)$ is 0 also, since $n_1, n_2 \in N = N^{\perp}$. So $B(n_1, n_1) = -\sqrt{a}B(n_1, n_2)$, implying that the F-part of $B(n_1, n_1)$ is 0 also. Hence $B(n_1, n_1) = 0$. This contradicts $(M,B)$ being anisotropic, unless $n_1 = n_2 = 0$. Thus $\gamma$ is $1 - 1$, and hence an isomorphism.

We must check that $\gamma$ is equivariant.

$$B'((n_1,n_2),(n_1',n_2')) = B(n_1,n_1') + aB(n_2,n_2') + [B(n_1,n_2') + B(n_1',n_2)]\sqrt{a}$$
$$= B(n_1 + \sqrt{a}n_2, n_1' + \sqrt{a}n_2')$$
$$= B(\gamma(n_1,n_2),\ \gamma(n_1',n_2')).$$

It follows that $(M,B) \sim (N \oplus N, B')$ as desired. □

<u>Lemma 1.4</u> $\ker I = \operatorname{im} S$.

<u>Step 1</u>. $\operatorname{im} S \subseteq \ker I$. Let $S[M,B] = [M, t \circ B]$.
Appling $I$ we obtain $[M \oplus M, (t \,\tilde{\circ}\, B), \tilde{\ell}]$. Let $N = \{(\sqrt{a}v, v) : v \in M\}$.
$N$ is $\tilde{\ell}$ invariant since $\tilde{\ell}(\sqrt{a}v, v) = (av, \sqrt{a}v)$. Further it is self-annihilating since

$$(t \,\tilde{\circ}\, B)\ ((\sqrt{a}v,v),\ (\sqrt{a}w,w)) = (t \circ B)\ ((\sqrt{a}v, \sqrt{a}w)) - a(t \circ B)\ (v,w)$$
$$= a(t \circ B)\ (v,w) - a(t \circ B)\ (v,w) = 0$$

Since $\operatorname{rank} N = \frac{1}{2} \operatorname{rank} (M \oplus M)$, $N = N^{\perp}$ and $M \oplus M \sim 0$.

<u>Step 2</u>. $\ker I \subseteq \operatorname{im} S$.

Let $(M,B)$ be anisotropic.

Suppose $I[M,B] = [M \oplus M, \tilde{B}, \tilde{\ell}] = 0$. Let $N$ be a metabolizer for $M \oplus M$.

Let $K = \{x \in M : (x,0) \in N\}$. If $x, y \in K$, $\tilde{B}((x,0),(y,0)) = B(x,y) = 0$. Since $(M,B)$ is anisotropic, $K = \{0\}$. Similarly, $\{x \in M: (0,x) \in N\} = 0$. Thus $N$ is the graph of a $1-1$ function $\ell: M \to M$. (We need $F$ a field to conclude that $\ell$ is an isomorphism of $M$ onto $M$.)

We may write $N = \{(\ell v, v), \ell: M \to M\}$.

$$\tilde{B}((\ell v,v), (\ell w,w)) = B(\ell v,\ell w) - aB(v,w) = 0$$

Thus $\ell$ is of degree a.

N is $\tilde{\ell}$ invariant, so that $(\ell v,v) \in N$ implies $(av,\ell v) \in N$. Hence $\ell(\ell v) = av$, and $\ell^2 = a$. Also

$$\tilde{B}((av,\ell v),(\ell w,w)) = B(av,\ell w) - aB(\ell v,w) = 0.$$

Thus $B(v,\ell w) = B(\ell v,w)$. Hence $\ell^* = \ell$.

N is already an F - vector space, with

$d(\ell v,v) = (d\ell(v),dv)$ for $d \in F$.

We make N into an $F(\sqrt{a})$ - vector space by defining

$$\sqrt{a}(\ell v,v) = (av,\ell v) = \tilde{\ell}(\ell v,v).$$

Now define $B_1$ on N by $B_1((\ell v,v), (\ell w,w)) = B(\ell v,w) + B(v,w)\sqrt{a}$. $B_1$ is symmetric since $B(v,\ell w) = B(\ell v,w)$. $B_1$ is clearly $F(\sqrt{a})$ - bilinear with the $F(\sqrt{a})$ - vector space structure defined above. $B_1$ is an inner product since B is.

Now we apply S to $[N,B_1]$, $S[N,B_1] = [N, t \circ B_1]$. Then define $\gamma: N \to M$ by $(\ell v,v) \to v$.

$$B(\gamma(\ell v,v), \gamma(\ell w,w)) = B(v,w) = (t \circ B_1)((\ell v,v), (\ell w,w)).$$

Hence $[N,t \circ B_1] = [M,B]$, and $[M,B] \in \text{im } S$. □

<u>Lemma 1.5</u> I <u>is</u> <u>onto</u>.

We define: d: $W^{+1}(-a,F;f) \to W^{-1}(F)$ by $[M,B,\ell] \to [M,\overline{B}]$ where $\overline{B}(x,y) = B(x,\ell^{-1}y)$.

$$\overline{B}(y,x) = B(y,\ell^{-1}x) = B(-\ell^{-1},y,x) = -B(x,\ell^{-1}y) = -\overline{B}(x,y).$$

Hence $\overline{B}$ is skew-symmetric. (This of course is the same d we have already defined. However, B symmetric yields $\overline{B}$ skew-symmetric.) Note that $W^{-1}(F) = 0$, so that kernel $d = W^{+1}(-a,F;f)$. )

Let $[M,B,\ell] \in W^{+1}(-a,F;f)$, so $[M,B,\ell] \in$ kernel d. Suppose $(M,B,\ell)$ is anisotropic. Since $[M,\overline{B}] = 0$, we let N be a metabolizer for $(M,\overline{B})$. Then $N \cap \ell N = \{0\}$ since $(M,B,\ell)$ is anisotropic.

Consider $[N,B_1] \in W(F)$, where $B_1 = B|_N$. Applying I we obtain $[N \oplus N,\tilde{B}_1,\tilde{\ell}]$. We shall show this is isomorphic to $(M,B,\ell)$. From this it follows that I is onto.

Define $\gamma: N \oplus N \to M$ by $(n_1,n_2) \to n_1 + \ell n_2$.

Then $B(\gamma(n_1,n_2),\gamma(n_1',n_2')) = B(n_1 + \ell n_2, n_1' + \ell n_2')$

$= B(n_1,n_1') + B(\ell n_2,\ell n_2') + B(n_1,\ell n_2') + B(\ell n_2,n_1')$

( $B(n_1,\ell n_2') = B(\ell n_2,n_1') = 0$ since $\overline{B} = 0$ on N. )

$= B(n_1,n_1') - aB(n_2,n_2'))$

$= \tilde{B}_1((n_1,n_2), (n_1',n_2'))$.

$\gamma\tilde{\ell}(n_1,n_2) = \gamma(an_2,n_1) = an_2 + \ell n_1$

$\ell\gamma(n_1,n_2) = \ell(n_1 + \ell n_2) = \ell n_1 + \ell^2 n_2 = \ell n_1 + an_2$.

Thus $\gamma$ is equivariant. Again, $\gamma$ is an isomorphism since $N \cap \ell N = 0$. Hence $(M,B,\ell) \simeq (N \oplus N,\tilde{B}_1,\tilde{\ell})$ as claimed. □

We have thus shown:

<u>Theorem 1.5</u> <u>There</u> <u>is</u> <u>an</u> <u>exact</u> <u>octagon</u>:

$$\begin{array}{ccccccc} & & W^{+1}(F(\sqrt{a})) & \overset{S}{\rightarrow} & W^{+1}(F) & \overset{I}{\rightarrow} & W^{+1}(-a,F;f) & & \\ & m \nearrow & & & & & & \searrow & \\ W^{+1}(F) & & & & & & & & 0 \\ & d \nwarrow & & & & & & \swarrow & \\ & & W^{-1}(-a,F;f) & \leftarrow & 0 & \leftarrow & 0 & & \end{array}$$

The map $S$ is Scharlau's transfer, with the other maps as previously described. □

The proof of Theorem 1.5 just given involved analyzing what actually is an 8 term exact sequence as we shall describe in Theorem 2.1. As one continues with Theorem 2.1, one should note the similarities in its proof to Theorem 1.5 In fact we may restrict the maps in Theorem 2.1 to obtain:

$$\begin{array}{ccccccc} & & W^{+1}(a,F;f) & \overset{S}{\rightarrow} & W^{+1}(a^2,F;g) & \overset{I}{\rightarrow} & W^{+1}(-a,F;f) & & \\ & m_1 \nearrow & & & & & & \searrow & \\ A(F;h) & & & & & & & & 0 \\ & d_{-1} \nwarrow & & & & & & \swarrow & \\ & & W^{-1}(-a,F;f) & \leftarrow & 0 & \leftarrow & 0 & & \end{array}$$

Here $f$ is the polynomial $f(t) = t^2 - a.$

$h$ is the polynomial $h(t) = t - 1$

$g$ is the polynomial $g(t) = t - a$

We then identify the groups:

$A(F;h) = W^{+1}(F)$ since $s = 1$ the identity map

$W^{+1}(a,F;f) = W(F(\sqrt{a}))$ by the trace lemma, since $\ell = \overline{\ell}$, and the induced - involution is trivial.

$W^{+1}(a^2,F;g) = W^{+1}(F)$ since $\ell = a$.

If one examines the induced maps under these "trace induced" identifications, then Theorem 2.1 will restrict to the Scharlau transfer as described.

2. <u>The exact octagon over a field.</u>

The Scharlau, Elman, Lam transfer sequence has been derived by computing successive kernels. For degree k mapping structures the maps of the Scharlau, Elman, Lam transfer sequence become:

$$d_\varepsilon: W^\varepsilon(-k,F) \to A(F)$$
$$[M,B,\ell] \to [M,\overline{B}] \quad \overline{B}(x,y) = k^{-1}B(x,\ell y)$$
$$m_\varepsilon: A(F) \to W^\varepsilon(k,F)$$
$$[M,B] \to [M \oplus M, B_\varepsilon, \ell_\varepsilon] \text{ where}$$
$$B_\varepsilon((x,y),(z,w)) = B(x,w) + \varepsilon B(z,y)$$
$$\ell_\varepsilon(x,y) = (\varepsilon k s^{-1} y, x) \qquad s \text{ is the symmetry operator for } B$$
$$S_\varepsilon: W^\varepsilon(k,F) \to W^\varepsilon(k^2,F)$$
$$[M,B,\ell] \to [M,B,\ell^2]$$
$$I_\varepsilon: W^\varepsilon(k^2,F) \to W^\varepsilon(-k,F)$$
$$[M,B,\ell] \to [M \oplus M, B \oplus -kB, \tilde{\ell}] \quad \tilde{\ell}(x,y) = (\ell y, x)$$

We shall prove:

Theorem 2.1 The above maps combine to yield an exact octagon:

$$\begin{array}{ccccccc}
 & & W^1(k,F) & \overset{S_1}{\to} & W^1(k^2,F) & \overset{I_1}{\to} & W^1(-k,F) & & \\
 & m_1 \nearrow & & & & & & \searrow d_1 & \\
A(F) & & & & & & & & A(F) \\
 & d_{-1} \nwarrow & & & & & & \swarrow m_{-1} & \\
 & & W^{-1}(-k,F) & \overset{I_{-1}}{\leftarrow} & W^{-1}(k^2,F) & \overset{S_{-1}}{\leftarrow} & W^{-1}(k,F) & &
\end{array}$$

We begin with the map $m_\varepsilon$. First, we observe that $\ell_\varepsilon$ is of degree $k$ with respect to $B_\varepsilon$ since:

$$B_\varepsilon(\ell_\varepsilon(x,y),\ \ell_\varepsilon(z,w)) = B_\varepsilon((\varepsilon ks^{-1}y,x),\ (\varepsilon ks^{-1}w,z))$$
$$= B\ (\varepsilon ks^{-1}y,z)\ +\ \not\varepsilon\ B(\not\varepsilon ks^{-1}w,x) = kB(s^{-1}w,x) + \varepsilon kB(s^{-1}y,x)$$
$$= kB(x,w) + \varepsilon kB(z,y) = k[B(x,w) + \varepsilon B(z,y)] = kB_\varepsilon((x,y),\ (z,w)).$$

$m_\varepsilon$ is well-defined, for if $(M,B) \sim 0$ has metabolizer $N$ then $m_\varepsilon(M,B)$ has metabolizer $N \oplus N$.

Lemma 2.2 $\ker S_\varepsilon = \operatorname{im} m_\varepsilon$.

Proof: Suppose $m_\varepsilon[M,B] = [M \oplus M,\ B_\varepsilon,\ \ell_\varepsilon]$. Then $M \oplus 0$ is an $\ell_\varepsilon^2$ invariant subspace, equal to its own annihilater by the way $B_\varepsilon$ is defined. Thus $\operatorname{im} m_\varepsilon \subseteq \ker S_\varepsilon$.

Conversely, suppose $[M,B,\ell] \in \ker S_\varepsilon$. Let $N$ be an $\ell^2$ invariant subspace of $M$ with $N = N^\perp$. $N$ is a metabolizer for

$S_\varepsilon[M,B,\ell] = [M,B,\ell^2]$. We assume as usual $(M,B,\ell)$ is anisotropic.

Note: $N \cap \ell N = 0$. This is seen as follows. Let $n \in N \cap \ell N$, and form $N_1 = \langle n, \ell n, \ldots, \ell^{w-1} n\rangle$ where $w$ is the degree of the minimal polynomial of $\ell$. Clearly $N_1$ is $\ell$ invariant, with $N_1 \subset N_1^\perp$. Thus, since we took $(M,B,\ell)$ anisotropic, $N_1 = 0$ and $n = 0$.

For vector spaces then it follows that $M = N \oplus \ell N$.

Define $B_1$ on $N$ by $B_1(n_1,n_2) = B(n_1,\ell n_2)$. We must verify that $B_1$ is an inner product. So consider the adjoint of $B_1$, $Ad_R B_1 \colon N \to Hom_F(N,F)$. Since we are working over a field, it suffices to show $Ad_R B_1$ is 1 - 1. We need $k \neq 0$, so that $\ell$ is non-singular. Suppose $B_1(-,n_2) = 0 = B(-,\ell n_2)$. Then $\ell n_2 \in N^\perp = N$. However, by the note, $N \cap \ell N = 0$. Thus $\ell n_2 = 0$. Since $\ell$ is non-singular $n_2 = 0$. Hence $B_1$ is non-singular.

We may thus form $[N,B_1] \in A(F)$. Applying $m_\ell$ we obtain $[N \oplus N, B_{1\,\varepsilon}, \ell_\varepsilon]$. Here

$$B_1(n_1,n_2) = B(n_1,\ell n_2) = \varepsilon B(\ell n_2,n_1) = \varepsilon k B(n_2,\ell^{-1} n_1)$$
$$= B(n_2,\ell(\varepsilon k \ell^{-2} n_1)) = B_1(n_2,\varepsilon k \ell^{-2} n_1)$$

Thus $s = \varepsilon k \ell^{-2}$ for $B_1$, or $\ell^2 = \varepsilon k s^{-1}$. It follows that

$$\ell_\varepsilon(x,y) = (\varepsilon k s^{-1} y, x) = (\ell^2 y, x)$$

Now define $\gamma \colon N \oplus N \to M$ by $(n_1,n_2) \to n_1 + \ell n_2$. Since $N \cap \ell N = 0$, $\gamma$ is an isomorphism. We now show that $\gamma$ is equivariant, ie. that $(N \oplus N, B_{1_\varepsilon}, \ell_\varepsilon) \cong (M,B,\ell)$. It follows that $\ker S_\varepsilon \subseteq \operatorname{im} m_\varepsilon$ as desired. We compute:

$$B_{1_\varepsilon}((x,y),(z,w)) = B_1(x,w) + \varepsilon B_1(z,y) = B(x,\ell w) + \varepsilon B(z,\ell y) +$$
$$B(x,z) + B(\ell y,\ell w) \quad (\text{since } B(x,z) = B(\ell y,\ell w) = 0 \text{ as } N = N^\perp)$$
$$= B(x,\ell w + z) + B(\ell y,\ell w + z)$$
$$= B(x + \ell y, z + \ell w) = B(\gamma(x,y), \gamma(z,w)).$$

Also, $\gamma(\ell_\varepsilon(n_1,n_2)) = \gamma(\varepsilon ks^{-1}n_2,n_1) = \gamma(\ell^2 n_2,n_1)$
$= \ell^2 n_2 + \ell n_1 = \ell(\ell n_2 + n_1) = \ell(\gamma(n_1,n_2))$. □

Lemma 2.3 $\ker I_\varepsilon = \operatorname{im} S_\varepsilon$.

Proof: Let $[M,B,\ell^2] \in \operatorname{im} S_\varepsilon$. We first show $I_\varepsilon([M,B,\ell^2]) = 0$. $I_\varepsilon[M,B,\ell] = [M \oplus M, B \oplus -kB, \tilde{\ell}^2]$, where $(\tilde{\ell}^2)(x,y) = (\ell^2 y,x)$.

Let $N = \{(x,\ell^{-1}x): x \in M\}$. $N$ is $(\tilde{\ell}^2)$ invariant. Further, $(B \oplus -kB)((x,\ell^{-1}x),(y,\ell^{-1}y)) = B(x,y) - kB(\ell^{-1}x,\ell^{-1}y)$
$= B(x,y) - B(x,y) = 0$.
Thus $N \subset N^\perp$, rank $N = (1/2)$ rank $M$, so that $N = N^\perp$ and $M \oplus M \sim 0$.

Conversely, let $[M,B,\ell] \in \ker I_\varepsilon$. So $I_\varepsilon[M,B,\ell] = [M \oplus M, B \oplus -kB, \tilde{\ell}] = 0$. Let $N$ be a metabolizer for $M \oplus M$. $N$ is $\tilde{\ell}$ invariant.

Now assume $(M,B,\ell)$ is anisotropic. We claim that $N$ is the graph, $N = \{(x,tx): x \in M\}$, of a $1-1$ function $t: M \to M$. Consider $K = \{x \in M: (x,0) \in N\}$. If $(x,0) \in N$, then $(0,x) \in N$ since $N$ is $\tilde{\ell}$ invariant. Thus $(\ell x,0) \in N$, and $\ell x \in K$. Hence $K$ is $\ell$ invariant. However, if $x,y \in K$, $B(x,y) = (B \oplus -kB)((x,0),(y,0)) = 0$, since $N = N^\perp$. Thus $K \subset K^\perp$. This contradicts $(M,B,\ell)$ anisotropic unless $K = \{0\}$. Similarly, $\{x \in M: (0,x) \in N\} = 0$. Thus $N$ is the graph of a $1-1$ function $t: M \to M$ and we can write $N = \{(x,tx): x \in M\}$. $t$ maps $M$ onto $M$ since $\dim N = \dim M$.

On $N$ we define $B_1,\ell_1$ by:

$B_1((x,tx),(y,ty)) = B(x,y)$.
$\ell_1(x,tx) = (\ell(tx),x)$ $\quad \ell_1: N \to N$ since $N$ is $\tilde{\ell}$ invariant.

We compute:

$$\begin{aligned} B_1(\ell_1(x,tx),\ell_1(y,ty)) &= B_1((\ell(tx),x),\ (\ell(ty),y)) \\ &= B(\ell(tx),\ell(ty)) \\ &= k^2B(tx,ty). \end{aligned}$$

However, $(B \oplus -kB)((x,tx),(y,ty)) = B(x,y) - kB(tx,ty) = 0$ since $N = N^\perp$. Thus, the above equals $kB(x,y) = kB_1((x,tx),(y,ty))$.

In order to have $[N,B_1,\ell_1] \in W^1(k,F)$, we still must show $B_1$ is an inner product. Again, it suffices to show that the adjoint is $1-1$.

$$\begin{aligned} Ad_R B_1 : N &\to Hom_F(N,F) \\ (x,tx) &\to B_1(-,(x,ty)) \end{aligned}$$

Suppose $B_1(-,(x,tx)) = 0$ on $N$. Then $B(-,x) = 0$ for all $y$, when $(y,ty) \in N$. Since we are over a field, $B(-,x) = 0$ on $M$. This contradicts $(M,B,\ell)$ anisotropic unless $x = 0$. Thus $Ad_R B_1$ is $1-1$, and hence an isomorphism.

We consider $[N,B_1,\ell_1]$. Define $\gamma:(N,B_1,\ell_1^2) \to (M,B,\ell)$ by $\gamma: (x,tx) \to x$. $\gamma$ is clearly an isomorphism. We claim $\gamma$ is equivariant, so that $S_\varepsilon[N,B_1,\ell_1] = [M,B,\ell]$ as desired.

$$B_1((x,tx),(y,ty)) = B(x,y) = B(\gamma(x,tx),\gamma(y,ty)).$$

$$\begin{aligned} \gamma \circ \ell_1^2(x,tx) &= \gamma \circ \ell_1(\ell(tx,x) = \gamma \circ (\ell x,\ell(tx)) \\ &= \ell x = \ell \circ \gamma(x,tx). \quad \square \end{aligned}$$

Lemma 2.4 ker $d_\varepsilon$ = im $I_\varepsilon$.

Let $I_\varepsilon[M,B,\ell] = [M \oplus M, B \oplus -kB, \tilde{\ell}]$. Applying $d_\varepsilon$ we obtain $[M \oplus M, \overline{B \oplus -kB}]$ where

$$\overline{(B \oplus -kB)}((x,y),(u,v)) = \frac{(B \oplus -kB)}{k}((x,y),\tilde{\ell}(u,v))$$

$$= \frac{1}{k}(B \oplus -kB)((x,y),(\ell v,u)).$$

In general, if $[M,B,\ell] \in W_\varepsilon(-k,F)$, applying $d_\varepsilon$ we obtain $[M,\overline{B}]$ where

$$\overline{B}(x,y) = \frac{1}{k} B(x,\ell y) = \frac{\varepsilon}{k} B(\ell y,x) = -\varepsilon B(y,\ell^{-1}x)$$

$$= \frac{1}{k} B(y,\ell(\ell^{-2}(-\varepsilon kx)) = \overline{B}(y,\ell^{-2}(-\varepsilon kx)).$$

So $s = -\varepsilon k\ell^{-2}$ is the symmetry operator for $\overline{B}$.

Now consider $N = M \oplus 0 \subset M \oplus M$. $N$ is $s = -\varepsilon k\tilde{\ell}^{-2}$ invariant rank $N = \frac{1}{2}$ rank $(M \oplus M)$ and

$$\overline{(B \oplus -kB)}((v,0),(w,0)) = \frac{1}{k}(B \oplus -kB)((v,0),(0,w)) = 0.$$

Thus $(M \oplus M, \overline{B \oplus -kB}) \sim 0$, with metabolizer $N$.

Conversely, suppose $[M,B,\ell] \in \ker d_\varepsilon$. Let $N$ be an $s = -\varepsilon k\ell^{-2}$ invariant subspace of $M$ with $N = N^\perp$ with respect to $\overline{B}$, where $\overline{B}(x,y) = \frac{1}{k}B(x,\ell y)$.

Define $\gamma\colon N \oplus N \to M$ by $(n_1,n_2) \to n_1 + \ell n_2$. We assume $(M,B,\ell)$ is anisotropic.

Claim: $\gamma$ is 1 - 1. Suppose $n_1 + \ell n_2 = 0$. Then form $K = \langle n_1, \ell n_1, \ldots, \ell^{r-1} n_1 \rangle$, where $r$ = degree of the minimal polynomial of $\ell$. Thus $K$ is $\ell$ invariant.

$$B(n_1,n_1) = B(n_1,-\ell n_2) = (-k)(\tfrac{1}{k}B(n_1,\ell n_2)) = -k\bar{B}(n_1,n_2) = 0$$

as $N = N^{\perp}$ with respect to $\bar{B}$. By similar computations, it follows that $K \subset K^{\perp}$. $K$ is $\ell$ invariant. This contradicts $(M,B,\ell)$ being anisotropic, unless $K = \{0\}$. Thus $\gamma$ is 1 - 1, and hence an isomorphism.

Consider $[N,B,\ell^2] \in W^{\varepsilon}(k^2,F)$. We must show $B$ is an inner product. Again, it suffices to show $\mathrm{Ad}_R B$ is 1 - 1. So suppose $B(-,n) = 0$ on $N$. Note that $B(\ell N,n) = 0$ since $N = N^{\perp}$. By the above $N \oplus \ell N = M$, so $B(-,n) = 0$ on $M$. Since $B$ is an inner product, $n = 0$ and $\mathrm{Ad}_R B$ is an isomorphism on $N$.

We now apply $I_{\varepsilon}$ to $[N,B,\ell^2]$. This yields $[N \oplus N,\ B \oplus -kB, \tilde{\ell}^2]$. Claim: $\gamma$ provides an equivariant isomorphism $\gamma\colon I_{\varepsilon}[N,B,\ell^2] \to [M,B,\ell]$. To see this, we compute:

$$\begin{aligned}(B \oplus -kB)((x,y),(u,v)) &= B(x,u) - kB(y,v)\\ &= B(x,u) + B(\ell y,\ell v) + B(x,\ell v) + B(\ell y,u)\\ &= B(x+\ell y,\ u+\ell v) = B(\gamma(x,y),\gamma(u,v))\end{aligned}$$

$$\gamma \circ \tilde{\ell}^2(x,y) = \gamma(\ell^2 y,x) = \ell^2 y + \ell x = \ell(\ell y + x) = \ell \circ \gamma(x,y). \quad \square$$

<u>Lemma 2.5</u> $\ker m_{\varepsilon} = \mathrm{im}\, d_{-\varepsilon}$.

Suppose $d_{-\varepsilon}[M,B,\ell] = [M,\overline{B}]$. The associated symmetry operator for $\overline{B}$ is $s = -(-\varepsilon)k\ell^{-2}$. Note: $B$ is $-\varepsilon$ symmetric.

Applying $m_\varepsilon$ to $[M,\overline{B}]$, we obtain $[M \oplus M,\overline{B}_\varepsilon,\ell_\varepsilon]$ where $\ell_\varepsilon(w_1,w_2) = (\varepsilon ks^{-1}w_2,w_1) = (\ell^2 w_2,w_1)$.

Consider $N \subset M \oplus M$ defined by $N = \{(x,\ell^{-1}x): x \in M\}$. $N$ is $\ell_\varepsilon$ invariant. Further

$$\begin{aligned}\overline{B}_\varepsilon((x,\ell^{-1}x),\ (y,\ell^{-1}y)) &= \overline{B}(x,\ell^{-1}y) + \varepsilon\overline{B}(y,\ell^{-1}x)\\ &= \tfrac{1}{k}B(x,y) + \varepsilon\tfrac{1}{k}B(y,x)\\ &= \tfrac{1}{k}B(x,y) + \not\varepsilon(-\not\varepsilon)\tfrac{1}{k}B(x,y) = 0.\end{aligned}$$

Thus $[M \oplus M,\overline{B}_\varepsilon,\ell_\varepsilon] = 0$ and $\operatorname{im} d_{-\varepsilon} \subseteq \ker m_\varepsilon$. Conversely, suppose $m_\varepsilon[M,B] = [M \oplus M,B_\varepsilon,\ell_\varepsilon] = 0$. Let $N$ be a metabolizer for $M \oplus M$ above. Assume $(M,B)$ is anisotropic.

Consider $K = \{x \in M: (x,0) \in N\}$. Since $N$ is $\ell_\varepsilon$ invariant $(x,0) \in N$ implies $(0,x) \in N$ and $(\varepsilon ks^{-1}x,0) \in N$. Also, $N$ is $\ell_\varepsilon^{-1}$ invariant since

$$B_\varepsilon((u,v),\ell_\varepsilon^{-1}(x,y)) = \tfrac{1}{k}B_\varepsilon(\ell_\varepsilon(u,v),(x,y)) = 0$$

for all $(u,v)$, $(x,y) \in N$. Thus $(x,0) \in N$ implies $(\varepsilon/k(sx),0) \in N$. $N$ is a subspace, so $(sx,0) \in N$ whenever $(x,0) \in N$. Thus $K$ is $s$ invariant. Further $B(x,y) = B_\varepsilon((x,0),(0,y)) = 0$ for all $x,y \in K$. This follows since $(y,0) \in N$ implies $\ell_\varepsilon(y,0) = (0,y) \in N$ also, and $N = N^\perp$ with respect to $B_\varepsilon$. Thus $K$ is an $s$ invariant, self annihilating subspace of $M$. This contradicts $M$ being anisotropic unless $K = 0$. Similarly $\{x \in M: (0,x) \in N\} = \{0\}$. It follows that

$N$ is the graph of a $1-1$ function $t: M \to M$, and we can write

$$N = \{(w,tw) : w \in M\}.$$

We now study this map $t$. First, since $N$ is $\ell_\varepsilon$ and $\ell_\varepsilon^{-1}$ invariant, if $(x,y) \in N$ then so is $\ell_\varepsilon(x,y) = (\varepsilon k s^{-1} y, x)$ and $\ell_\varepsilon^{-1}(x,y) = (y, \varepsilon/k\ sx)$.

Thus, if $(y,ty) \in N$, so is $(ty, \varepsilon/k\ sy)$, so that $t(ty) = \frac{\varepsilon}{k} sy$. More simply, $t^2 = \frac{\varepsilon}{k} s$, or $\varepsilon k t^2 = s$.

Moreover, $B_\varepsilon((x,tx),(y,ty)) = 0$

$$\begin{aligned}
&= B(x,ty) + \varepsilon B(y,tx) = B(ty,sx) + \varepsilon B(y,tx) \\
&= B(y,t^*sx) + \varepsilon B(y,tx) \qquad \text{where } t^* = \text{adjoint of } t \\
&= B(y,(t^*s + \varepsilon t)x) = 0
\end{aligned}$$

Since $B$ is non-singular, $t^*s + \varepsilon t = 0$. Thus

$$t^* = -\varepsilon t s^{-1} = (-\varepsilon t)(\varepsilon/k\ t^{-2}) = \frac{-t^{-1}}{k}.$$

On $M$, define an inner product $B_1$ by $B_1(x,y) = kB(x,ty)$. $B_1$ is non-singular since $B$ and $t$ are, as usual.

$B_1$ is $(-\varepsilon)$ symmetric since:

$$\begin{aligned}
B_1(x,y) &= kB(x,ty) = kB(ty,sx) = kB(y,t^*sx) \\
&= kB(y,-\varepsilon tx) = -\varepsilon kB(y,tx) = -\varepsilon B_1(y,x).
\end{aligned}$$

Now consider $(M,B_1,t^{-1})$. $B_1(t^{-1}x,t^{-1}y) = kB(t^{-1}x,y) = kB(x,t^{*-1}y)$

$$\begin{aligned}
&= kB(x,-kty) = (-k)(k)B(x,ty) \\
&= (-k)B_1(x,y)
\end{aligned}$$

Thus $[M,B_1,t^{-1}] \in W^{-\varepsilon}(-k,F)$. Applying $d_{-\varepsilon}$ we obtain $[M,\bar{B}_1]$.

$\bar{B}_1(x,y) = k^{-1}B_1(x,t^{-1}y) = B(x,y)$ as desired. □

This completes the proof of the exact octagon over a field F, when $k \neq 0$. The failure of this proof for a Dedekind domain, eg. Z the ring of integers, is the verification that the bilinear maps we are constructing are actually Z-inner products. In order to overcome this difficulty, we use a different approach. Namely, we study a boundary sequence relating $W(k,Z)$ to $W(k,Q)$. This boundary sequence then enables us to study the octagon over Z.

## Chapter VI THE BOUNDARY

Our ultimate goal is to study the octagon over $Z$, when $k = \pm 1$. In order to do this, we relate $W(k,Z)$ to $W(k,Q)$ by means of an exact sequence. In Section 1, the boundary homomorphism, $\partial$, is defined. This enables us to establish an exact sequence:

$$0 \rightarrow W(k,Z) \rightarrow \mathcal{W}(k,Q) \xrightarrow{\partial} W(k,Q/Z).$$

Next let $S = Z[t,t^{-1}]/(f(t))$, where $f(t)$ is a $T_k$ fixed monic, integral, irreducible polynomial. We have the decomposition $\mathcal{W}(k,Q) \simeq \oplus W(k,Q;f) = \oplus W(k,Q;S)$. It is thus natural to consider the restriction of $\partial$ to $W(k,Q;S)$. We denote this restriction $\partial(S)$. We wish to compute $\partial(S)$. This will eventually allow us to analyze $\partial$.

The first step is the reduction of the study of $\partial(S)$ to the study of $\partial(D): W(k,Q;D) \rightarrow W(k,Q/Z;D)$ where $D$ is the Dedekind ring of integers in $E = Q[t,t^{-1}]/(f(t))$. This is done in Section 2.

In Section 3, we begin the computation of $\partial$ by studying the local case $\partial(D,P)$. Here $\partial(D,P)$ is the localization of the $\partial(D)$ sequence at a - invariant maximal ideal $P$ in $D$. We use this in Section 4 to compute the cokernel of $\partial(D)$.

### 1. <u>The boundary homomorphism</u>

We shall construct an exact sequence

$$0 \rightarrow W(k,Z) \xrightarrow{i} \mathcal{W}(k,Q) \xrightarrow{\partial} W(k,Q/Z).$$

The script $\mathcal{W}$ indicates that we have placed a restriction on the minimal polynomial of $\ell$ in the degree $k$ mapping structure $[M,B,\ell]$. This restriction is that the minimal polynomial of $\ell$ be a monic integral polynomial.

We also should observe that this construction works equally well for $Z$ replaced by $D$ a Dedekind Domain, $Q$ replaced by $E$ the quotient field of $D$, and $Q/Z$ replaced by $E/D$. The sequence then reads:

$$0 \to W(k,D) \overset{i}{\to} \mathcal{W}(k,E) \overset{\partial}{\to} W(k,E/D).$$

Note that this construction also applies to the asymmetric case. We similarly will obtain an exact sequence:

$$0 \to A(Z) \to \mathcal{A}(Q) \to A(Q/Z).$$

If $[M,B] \in A(Q)$, with symmetry operator $s$ satisfying $B(x,y) = B(y,sx)$, we then require that the minimal polynomial of $s$ be a monic integral polynomial in order for $[M,B]$ to be in script $\mathcal{A}(Q)$.

The restriction that we have placed on the minimal polynomial of $\ell$ (respectively $s$) is dictated by:

<u>Proposition 1.1</u> <u>A degree</u> $k$ <u>mapping structure</u> $(M,B,\ell)$ <u>over</u> $Q$ <u>contains an</u> $\ell$ <u>invariant integral lattice</u> $A$ <u>if and only if the minimal polynomial of</u> $\ell$ <u>is a monic, integral polynomial.</u>

<u>Comments</u>: (1) $\ell$ is replaced by the symmetry operator $s$ for the asymmetric case, $A(Q)$.

(2) All lattices are assumed to be full. [B,S 99] This means that $A$ is a finitely generated $Z$-submodule of $M$ with rank $A$ = rank $M$.

(3) An integral lattice $A$ is one on which the inner product $B$ is integrally valued.

<u>Proof</u>: Necessity. We assume $M$ contains an $\ell$ invariant integral lattice $A$. Since $B|_A$ is integral the characteristic polynomial of $\ell$ is integral. Since the characteristic polynomial of $\ell$ is integral, so is the minimal polynomial [L - 1 402].

Sufficiency. Conversely, suppose that the minimal polynomial, $f(x)$, of $\ell$ is integral. Write

$$f(x) = a_0 + a_1 x + \dots + a_{m-1}x^{m-1} + x^m.$$

By Lemma II. 1.1, $f(\ell) = f(\ell^*) = 0$. Thus, we obtain the identities

$$\ell^m = -\sum_{i=0}^{m-1} a_i \ell^i, \quad (\ell^*)^m = -\sum_{i=0}^{m-1} a_i (\ell^*)^i.$$

We now construct an $\ell$, $\ell^*$ invariant integral lattice. Let $\{e_1, \dots, e_n\}$ be a basis for $M$ over $Q$. Since $B(x,y) \in Q$ for $x,y \in M$, we can find integers $r_{ijk}$, $s_{ijk}$ so that $r_{ijk}B(\ell^k e_i, e_j) \in Z$ and $s_{ijk}B((\ell^*)^k e_i, e_j) \in Z$ for $i,j \leq n$ and $k < m$. Let $d = \Pi r_{ijk}s_{ijk}$. Then clearly $\{de_1, \dots, de_n\}$ generates a free $Z$-module on which $B$ is integrally valued.

Let $f_i = de_i$. We then define $A$ to be the $Z$-lattice spanned by

$$\{f_1, \dots, f_n, \ell f_1, \ell f_2, \dots, \ell f_n, \ell^2 f_1, \dots, \ell^2 f_n, \dots$$
$$\ell^{m-1} f_1, \dots, \ell^{m-1} f_n, \ell^* f_1, \dots \ell^* f_n, \dots \ell^{*m-1} f_1, \dots \ell^{*m-1} f_n\}.$$

Here $m = \text{degree } f(x)$, the minimal polynomial of $\ell$ and $\ell^*$. The identities for $\ell^m$, and $(\ell^*)^m$, together with $\ell\ell^* = k$ show that $A$ is $\ell$ and $\ell^*$ invariant. Since $B$ is integral on $\{f_1, \dots, f_n\}$. it follows that $B$ is integral on $A$ because $\ell$ has degree $k$. Thus $A$ is an $\ell$, $\ell^*$ invariant lattice as desired. □

<u>Definition 1.2</u> $\mathcal{W}(k,Q)$ (<u>respectively</u> $\mathcal{A}(Q)$) <u>denotes Witt equivalence classes</u> $[M,B,\ell]$ <u>in</u> $W(k,Q)$ <u>for which the minimal polynomial of</u> $\ell$ (<u>respectively</u> $s$) <u>is integral</u>.

We now define the maps in the boundary sequence:

$$0 \to W(k,Z) \xrightarrow{i} \mathcal{W}(k,Q) \xrightarrow{\partial} W(k,Q/Z).$$

If $[M,B,\ell] \in W(k,Z)$, $i[M,B,\ell] \equiv [M,B,\ell] \otimes_Z Q$.

The boundary homomorphism $\partial$ is more involved. Let $[M,B,\ell] \in \mathcal{W}(k,Q)$. By Proposition 1.1, we can find an $\ell$ invariant integral lattice $L \subset M$. We define the dual lattice of $L$ to be

$$L^\# = \{x \in M: B(x,L) \subset Z\}.$$

We observe that $L^\# \supset L$, with rank $L^\# = \text{rank } L$. Thus $L^\#/L$ is a finitely generated torsion $Z$-module.

If $\overline{x},\overline{y} \in L^{\#}/L$, we let $x$ and $y$ be preimages of these equivalence classes in $L^{\#}$. Let $q\colon Q \to Q/Z$ be the quotient map. We define $B'\colon L^{\#}/L \times L^{\#}/L \to Q/Z$ by:

$$B'(\overline{x},\overline{y}) = q \circ B(x,y).$$

We define $\ell'\colon L^{\#}/L \to L^{\#}/L$ by:

$$\ell'(\overline{x}) = \overline{\ell(x)}$$

We now need to show that $[L^{\#}/L,B',\ell']$ is an inner product space in $W(k,Q/Z)$. Once this is done, we define $\partial$ by:

$$\partial\colon [M,B,\ell] \to [L^{\#}/L,B',\ell'].$$

To begin with, since $L$ is $\ell$ invariant so is $L^{\#}$. Thus $\ell'$ is well-defined.

The fact that $B$ is integral on $L$ implies that $B'$ is well-defined on $L^{\#}/L$, with values in $Q/Z$.

We must show that $B'$ is non-singular. In order to do this we define

$$h\colon L^{\#} \to \mathrm{Hom}_Z(L^{\#}/L,Q/Z)$$

by

$$x \to B'(\overline{x},-)$$

We need to show that $h$ is epic, with kernel $L$. We shall need:

<u>Lemma 1.3</u> <u>For a lattice</u> $L$, $(L^{\#})^{\#} = L$.

<u>Proof</u>: If $\{v_1,\ldots,v_n\}$ is a basis for $L$, then a basis for $L^{\#}$ is given by $\{v_1^{\#},\ldots,v_n^{\#}\}$, where $B(v_i,v_j^{\#}) = \delta_{ij}$. It follows that a basis for $(L^{\#})^{\#}$ is $\{v_1,\ldots,v_n\}$. □

We show $h$ is onto. Let $g: L^{\#}/L \to Q/Z$ be given. Let $\{w_1,\ldots,w_n\}$ be a basis for $L^{\#}$. Then define $\hat{g}:L^{\#} \to Q$ as follows. $\hat{g}(w_i)$ is chosen so that:

$$q \circ \hat{g}(w_i) = g(\overline{w}_i) \qquad \overline{w}_i \in L^{\#}/L$$

Then extend $\hat{g}$ linearly to $L^{\#}$.

In fact, tensoring with $Q$, we may assume $\hat{g} \in \mathrm{Hom}_Z(M,Q)$. Thus since $B$ is an inner product we may write

$$B(v_1,-) = \hat{g}(-).$$

However $B(v_1,-)$ is integer valued on $L$ since $\hat{g}(-) = B(v_1,-)$ and $\hat{g}$ is integral on $L$. Thus $v_1 \in L^{\#}$. It is now clear that

$$g(-) = q \circ \hat{g}(-) = q \circ B(v_1,-) = B'(v_1,-)$$

and $h$ is onto.

In order to show that the kernel of $h$ is $L$, suppose $h(w) = 0$. So $B'(\overline{w},-) = 0$. Thus for all $v \in L^{\#}$, we have $B(w,v) \in Z$. Hence, $w \in (L^{\#})^{\#} = L$ by Lemma 1.3, and the kernel of $h$ is $L$.

Thus, given $[M,B,\ell] \in \mathcal{W}(k,Q)$, we have a method for obtaining an element in $W(k,Q/Z)$. We define:

$$\partial[M,B,\ell] = [L^{\#}/L,B',\ell']$$

<u>Lemma 1.4</u> $\partial$ <u>is well-defined</u>.

In order to show that $\partial$ is well-defined, we must show that this construction is independent of the lattice chosen, and that $\partial$ is trivial on metabolic forms. If we do this, then $\partial$ preserves Witt-equivalence and gives a homomorphism of Witt groups.

(a) Independence of lattice.

Let $L_\circ$ be another choice of $\ell$ invariant integral lattice. Without loss of generality, $L_\circ \subset L$, for otherwise, we can show that both $L_\circ$ and $L$ give the same result as the $\ell$ invariant integral lattice $L_\circ \cap L$.

We then have $L_\circ \subset L \subset L^{\#} \subset L_\circ^{\#}$. Consider $L/L_\circ \subset L_\circ^{\#}/L_\circ$. The annihilator of $L/L_\circ$ in $L_\circ^{\#}/L_\circ$ is $(L/L_\circ)^{\perp} = L^{\#}/L_\circ$. Thus, by I.6.4, $[L_\circ^{\#}/L_\circ, B_\circ', \ell_\circ']$ is Witt -equivalent to $[(L/L_\circ)^{\perp}/(L/L_\circ), B_\circ'', \ell_\circ'']$ where $B_\circ''$, $\ell_\circ''$ indicate the appropriate induced forms. However,

$$(L/L_\circ)^{\perp}/(L/L_\circ) = (L^{\#}/L_\circ)\ /(L/L_\circ) \simeq L^{\#}/L.$$

Hence $[L_\circ^{\#}/L_\circ, B_\circ', \ell_\circ'] = [L^{\#}/L,B', \ell']$ as was to be shown.

(b) We now show $\partial(\text{metabolic}) \sim 0$. So let $(M,B,\ell) \sim 0$ with metabolizer $N$. Let $L$ be an $\ell$ invariant lattice. Define $N_1 = N \cap L^{\#}$.

Clearly $N_1$ is $\ell$ invariant. We also have the exact sequence $0 \to N_1 \to L^{\#} \to L^{\#}/N_1 \to 0$.

Claim: This sequence splits so that $N_1$ is a summand of $L^{\#}$.

To show this, it clearly suffices to show $L^{\#}/N_1$ is torsion free and hence projective. Suppose to the contrary that there is

$$x \in L^{\#},\ x \notin N_1,\ d \neq 0 \text{ with } dx \in N_1.$$

Now $dx \in N_1 \subset N$, so

$$B(dx,y) = dB(x,y) = 0 \quad \text{for all} \quad y \in N_1.$$

Tensoring with $Q$, $(N_1 \cap L^{\#}) \otimes_Z Q \supset N$.
Therefore

$$dB(x,y) = 0 \qquad \text{for all} \quad y \in N, \text{ and}$$
$$B(x,y) = 0 \qquad \text{for } y \in N.$$

Hence

$$x \in N^{\perp} = N, \qquad x \in L^{\#}$$

so that

$$x \in N \cap L^{\#} = N_1. \text{ Contradiction.}$$

It follows that $L^{\#}/N_1$ is torsion free, and $N_1$ is a summand of $L^{\#}$.

Now let $H = (L^{\#} \cap N)/L \subset L^{\#}/L$. Clearly $H \subset H^{\perp}$.

Conversely, suppose $\bar{k} \in H^{\perp}$. This means $B'(\bar{k},h) \in Z$ for all $h \in H$. Let $k$ be a lift of $\bar{k}$ to $L^{\#}$. Consider the diagram

$$\begin{array}{ccc} B(k,-) : L^{\#} \cap N & \to & Z \\ \downarrow & \nearrow & \\ L^{\#} & & f \end{array}$$

$B(k,-)$ extends to $L^{\#}$ since $L^{\#} \cap N = N_1$ is a summand. Call this extension f. Then $f(-) = B(w,-)$ since B is non-singular. $B(w,x) \in Z$ for $x \in L^{\#}$. Thus $w \in (L^{\#})^{\#} = L$. Now consider $k - w$. $B(k-w,x) = 0$ for $x \in L^{\#} \cap N = N_1$. Thus $k - w \in (L^{\#} \cap N)^{\perp}$. Since $L^{\#}$ is a lattice, $k - w \in N^{\perp} = N$. Clearly $k - w \in L^{\#}$. Thus $k - w \in L^{\#} \cap N = N_1$.

Now $(\overline{k - w}) \in H$. However, $w \in L$, so $\overline{w} = 0 \in H$. Thus $\overline{k} \in H$ as desired. So $H^{\perp} = H$ and $\partial(\text{metabolic}) \sim 0$. $\square$

Since we have now shown that $\partial$ is well-defined and independent of the choices made, we are ready to prove:

<u>Theorem 1.5</u> <u>The sequence</u>

$$0 \to W(k,Z) \xrightarrow{i} \mathcal{W}(k,Q) \xrightarrow{\partial} W(k,Q/Z)$$

<u>is exact</u>.

<u>Proof</u>: i is 1 - 1 by Lemma I 5.4.

im i $\subseteq$ ker $\partial$.

Let $[M,B,\ell] \to [M,B,\ell] \otimes_Z Q$. Choose the lattice $M \otimes 1 = L$ in $M \otimes_Z Q$. For this choice of L, we have $L \cong \mathrm{Hom}_Z(L,Z)$ since B is non-singular on M. Thus $L^{\#} = L$ and $\partial \circ i = 0$.

ker $\partial \subseteq$ im i.

Suppose $\partial([M,B,\ell]) = 0$. Let $H \subset L^{\#}/L$ be a metabolizer for $(L^{\#}/L, B', \ell')$. Let $L_0$ = inverse image of H in $L^{\#} \subset M$ under the projection $L^{\#} \to L^{\#}/L$. Then $B|_{L_o}$ has values in $Z$ since $H$ is a self-annihilating subspace of $L^{\#}/L$, meaning that $B' = 0 \in Q/Z$ on $H$.

$L_o^{\#} = \{x \in M: B(x,L_o) \subset Z\}$. If $x \in L_o^{\#}$, projecting to $L^{\#}/L$, $\bar{x} \in H^{\perp} = H$. Thus $x \in L_o$. Obviously $L_o \subset L_o^{\#}$, so $L_o = L_o^{\#} \cong \mathrm{Hom}_Z(L_o,Z)$, and the adjoint is an isomorphism on $L_o$.

$L_o$ is $\ell$ invariant since $H$ is $\ell'$ invariant. Thus consider $[L_o, B|_{L_o}, \ell|_{L_o}] \in W(k,Z)$. Applying $i$, we obtain $[M,B,\ell]$, since $L$ is a full lattice. □

Corollary 1.6 The sequence

$$0 \to A(Z) \to A(Q) \to A(Q/Z) \quad \text{is exact.} \quad \square$$

## 2. Reducing to the maximal order

We continue our study of the boundary homomorphism by recalling the computation of $W(k,Q)$ and $A(Q)$. This was done as follows. Let

$$S = Z(\theta) = Z[t,t^{-1}]/(f(t)),$$

where $f(t)$ is a monic, integral, irreducible $T_k$ fixed polynomial. Let $E = Q[t,t^{-1}]/(f(t))$. Then, as we have seen, the field $E$ has an involution $-$ induced from $T_k$. The fixed field of $-$ is denoted by $F$. $S$ is an order in $O(E)$, the ring of integers in $E$. To simplify our notation, we write $O(E) = D$.

As in Theorem IV 1.8, we have the following computation:

$$W(k,Q) \approx \bigoplus_{f \in \mathcal{B}} W(k,Q;f) \approx \bigoplus W(k,Q;S).$$

Note: Here we are using the symbol $\mathcal{B}$ to denote the collection of $T_k$ fixed, monic, integral, irreducible polynomials. This should not be confused with IV 1.8 where the integral requirement is omitted, for $W(k,Q)$.

We denote the restriction of $\partial$ to $W(k,Q;S)$ by $\partial(S)$.

Lemma 2.1 There are exact sequences:

$$0 \to W(k,Z;S) \xrightarrow{i} W(k,Q;S) \xrightarrow{\partial(S)} W(k,Q/Z;S)$$

$$0 \to A(Z;S) \xrightarrow{i} A(Q;S) \xrightarrow{\partial(S)} A(Q/Z;S).$$

These follow from Theorem 1.5, since the S-module structure is preserved by $i$ and $\partial(S)$. □

In order to study and compute the map $\partial(S)$ we begin by comparing this with the exact sequence for $\partial(D)$ where $D$ is the maximal order. We have the following commutative diagram of forgetful maps.

Lemma 2.2 Let $f_1, f_2, f_3$ be the maps which forget the D-module structure and remember only the S-module structure. Then the diagram below commutes:

$$\begin{array}{ccccccc} 0 & \to & W(k,Z;D) & \xrightarrow{i_1} & W(k,Q;D) & \xrightarrow{\partial(D)} & W(k,Q/Z;D) \\ & & \downarrow f_1 & & \downarrow f_2 & & \downarrow f_3 \\ 0 & \to & W(k,Z;S) & \xrightarrow{i_2} & W(k,Q;S) & \xrightarrow{\partial(S)} & W(k,Q/Z;S) \end{array}$$

Proof: We begin by remarking that the notation $W(k,Q;D)$ or $W(k,Q;S)$ is somewhat redundant. Indeed $W(k,Q;D) \simeq W(k,Q;S) \simeq H(E)$.

In order to verify commutativity recall that $\partial(D)$ is defined by choosing an integral D-lattice, L, forming $L^{\#}/L$, etc. However, "by forgetting," the D-lattice L is also an S-lattice, $L^{\#}$ is the same, and so consequently is $L^{\#}/L$. □

Thus, we see that we can reduce the study of $\partial(S)$ to studying $\partial(D)$ where D is the maximal order. However, this reduction is more complicated than might first appear since we still must compute both $\partial(D)$ and the maps $f_i$.

To see precisely the problems involved in this reduction we continue with the study of $\partial(D)$ for D the maximal order. We find it profitable to give the computation for $\partial(D)$ on the "Hermitian" level. Thus we recall the following identifications.

$W(k,Z;D) \simeq H(\Delta^{-1}(D/Z))$ where $\Delta^{-1}(D/Z)$ is the inverse different of D over Z

$W(k,Q;D) \simeq H(E)$

$$W(k,Q/Z;D) \simeq H(E/\Delta^{-1}(D/Z)) \simeq \bigoplus_{\mathcal{P} = \overline{\mathcal{P}}} H(D/\mathcal{P})$$

In the asymmetric case we likewise had:

$$A(Z;D) \approx H_\theta(\Delta^{-1}(D/Z))$$

$$A(Q;D) \approx H_\theta(E)$$

$$A(Q/Z;D) \approx H_\theta(E/\Delta^{-1}(D/Z))$$

These isomorphisms together with Lemma 2.1 lead us to the exact sequence:

$$0 \to H_u(I) \xrightarrow{i} H_u(E) \xrightarrow{\partial(D)} H_u(E/I)$$

where $I = \Delta^{-1}(D/Z)$.
We should discuss this new boundary map which we continue to call $\partial(D)$ since there should be no confusion.

Let $[M,B] \in H_u(E)$. We first must construct a D-lattice $L$ in $M$ so that $B|_L$ takes values in $I$. In order to do this we follow Proposition 1.1. Let $\{e_1,\ldots,e_n\}$ be a basis for $M$. Then we can find non-zero integers $r_{ijk}, s_{ijk}$ so that $r_{ijk}B(\theta^k e_i, e_j) \in I$ and $s_{ijk}B(\overline{\theta}e_i, e_j) \in I$ for $i,j \leq n$ and $k < m$, where $m$ is the degree of the minimal polynomial of $\theta$. Let $d = \Pi r_{ijk}s_{ijk}$. Then the lattice $L$ generated by $\{de_1,\ldots,de_n,\ldots,\theta^m de_1,\ldots\theta^m de_n\}$ is a D-lattice on which $B$ is I-valued as in 1.1.

Let $L^\# = \{v \in M: B(v,L) \subset I\}$. $L^\#$ is also a D-lattice. We form $L^\#/L$ with induced E/I-valued inner product $B'$ as before. Then define $\partial(D)$: $[M,B] \to [L^\#/L, B']$. With $\partial(D)$ so defined, the following is clear.

<u>Proposition 2.3</u> $\partial(D)$ <u>is well-defined for Hermitian and the following diagram commutes</u>

$$\begin{array}{ccccccc} 0 & \to & H(\Delta^{-1}(D/Z)) & \overset{i}{\to} & H(E) & \overset{\partial(D)}{\to} & H(E/\Delta^{-1}(D/Z)) \\ & & \downarrow \simeq t & & \downarrow \simeq t & & \downarrow \simeq t \\ 0 & \to & W(k,Z;D) & \overset{i}{\to} & W(k,Q;D) & \overset{\partial(D)}{\to} & W(k,Q/Z;D) \end{array}$$

The vertical isomorphisms are induced by trace, t, of E over Q. □

So the method we shall employ for computing $\partial(D)$ is to study the corresponding boundary for Hermitian. The image of boundary is the group $H(E/\Delta^{-1}(D/Z)) \simeq W(k,Q/Z;D)$. We recall the computation of this group given in Chapter IV 3.5, 3.8. We had:

$$W(k,Q/Z;D) \overset{g}{\to} \bigoplus_{P=\bar{P}} W(k,F_p;D/P) \overset{tr}{\leftarrow} \bigoplus_{P=\bar{P}} H(D/P).$$

Here we sum over all - invariant maximal ideals in D. The isomorphism g is induced by selecting a generator, say $\frac{1}{p}$, for the p-torsion in Q/Z. tr denotes the isomorphism induced by trace on finite fields, $D/P$ over $F_p$.

Similar computations apply to W(k,Q/Z;S). We use the letter $M$ to denote a - invariant maximal ideal in S.

<u>Proposition 2.4</u> <u>The diagram below commutes.</u>

$$\begin{array}{ccccc} W(k,Q/Z;D) & \overset{g}{\to} & \bigoplus_{P=\bar{P}} W(k,F_p;D/P) & \overset{tr}{\leftarrow} & \bigoplus_{P=\bar{P}} H(D/P) \\ \downarrow f_3 & & & & \downarrow \oplus tr \\ W(k,Q/Z;S) & \overset{g}{\to} & \bigoplus_{M=\bar{M}} W(k,F_p;S/M) & \overset{tr}{\leftarrow} & \bigoplus_{M=\bar{M}} H(S/M) \end{array}$$

The map tr: $H(D/P) \to H(S/M)$ where $M = P \cap S$ is given by trace on finite fields.

Proof: Let $[M,B] \in H(D/P)$. Then apply tr to obtain an element in $W(k,F_p;D/P)$. Apply the isomorphism $g^{-1}$ and "forget" to obtain an element in $W(k,Q/Z;S)$.

However, tr: $D/P \to F_p$ is the same as the composition tr: $D/P \to S/P\cap S \to F_p$. Thus $f_3 \circ g^{-1} \circ tr = g^{-1} \circ tr \circ \oplus$ tr, ie. the above diagram commutes. □

In order to appreciate Proposition 2.4, we collect those - invariant maximal ideals in D which lie over a given - invariant maximal ideal in S. Let $T(M) = \{P: P$ is a - invariant maximal ideal in D with $P \cap S = M\}$, where $M$ is a - invariant maximal ideal in S. $T(M)$ may or may not be empty. We shall discuss $T(M)$ further in Chapter VII 4.

We now define the local boundary by:

$\partial(D,P)$: $W(k,Q;D) \to H(D/P)$ by $\partial(D,P) = q(P) \circ \partial(D)$ where

$q(P)$ is the composition of $(tr^{-1} \circ g)$ with projection to the $P^{th}$ coordinate in $\oplus H(D/P)$.

The local boundary $\partial(S,M)$ is similarly defined.

Theorem 2.5 $\partial(S,M) = \bigoplus_{P \in T(M)} tr \circ \partial(D,P)$

Proof: By Lemma 2.2, $f_3 \circ \partial(D) = \partial(S)$. Thus, using Proposition 2.4 we need only determine which - invariant maximal ideals $P$ in D lie over $M$ in $S$. These are given by $T(M)$ by definition. Thus, the commutative diagram in 2.4 yields:

$$\partial(S,M) = \bigoplus_{P \in T(M)} \mathrm{tr} \circ q(P) \circ \partial(D)$$

$$= \bigoplus_{P \in T(M)} \mathrm{tr} \circ \partial(D,P)$$

In order to tabulate the progress we have made and to indicate the problems involved in the computation of $\partial(S)$ we record the following diagram.

$$\begin{array}{ccccccccc}
0 \to H(\Delta^{-1}(D/Z)) & \xrightarrow{i} & H(E) & \xrightarrow{\partial(D)} & H(E/\Delta^{-1}(D/Z)) & \searrow & & & \\
\downarrow t & & \downarrow t & & \downarrow t & & & & \\
0 \to W(k,Z;D) & \xrightarrow{i} & W(k,Q;D) & \xrightarrow{\partial(D)} & W(k,Q/Z;D) & \xrightarrow{g} & \bigoplus_{P = \bar{P}} W(k,F_p;D/P) & \cong & \bigoplus H(D/P) \\
\downarrow f_1 & & \downarrow f_2 & & \downarrow f_3 & & & & \downarrow \mathrm{tr} \\
0 \to W(k,Z;S) & \xrightarrow{i} & W(k,Q;S) & \xrightarrow{\partial(S)} & W(k,Q/Z;S) & \xrightarrow{g} & \bigoplus_{M = \bar{M}} W(k,F_p;S/M) & \cong & \bigoplus H(S/M)
\end{array}$$

We wish to compute the group $W(k,Z;S)$. The idea is to use the above diagram. We compute the group $W(k,Z;D)$ for the maximal order $D$ by using the exact sequence with $\partial(D)$. Then we compute the map $f_1$. We first discuss $f_1$.

Lemma 2.6 $f_1$ is 1 - 1.

Proof: Clear by diagram chase since $i$ is 1 - 1 and $f_2$ is an isomorphism. □

In order to compute the cokernel of $f_1$ we recall some homological algebra. [M 50] $\operatorname{im} \partial(D) \cap \ker f_3 = \operatorname{coker} f_1$. Thus we need to compute $\partial(D)$ and $f_3$.

By Proposition 2.4, $f_3$ is determined by two things:

(1) The map $tr: H(D/P) \to H(S/P \cap S)$ induced by trace on finite fields.

(2) $T(M) = \{$maximal ideals $P$ in $D$: $P \cap S = M$, $P = \overline{P}\}$

We shall discuss both of these in Chapter VII.

In analyzing the group $W(k,Z;D) \cong H(\Delta^{-1}(D/Z))$ and $\partial(D)$ we use the isomorphic Hermitian sequence.

$$0 \to H(\Delta^{-1}(D/Z)) \to H(E) \xrightarrow{\partial(D)} H(E/\Delta^{-1}(D/Z)) \cong \bigoplus_{P=\overline{P}} H(D/P)$$

Again we call $\partial(D,P) = q(P) \circ \partial(D)$, where $q(P)$ denotes projection to the $P^{th}$ component $H(D/P)$. We may identify $\partial(D)$ with $\bigoplus_P \partial(D,P)$, so that in essence our task is to compute $\partial(D,P)$. It is at this point that we must be very careful however.

The isomorphism $H(E/\Delta^{-1}(D/Z)) \cong \bigoplus_{P=\overline{P}} H(D/P)$ can be given in any fashion and will enable us to compute $H(\Delta^{-1}(D/Z)) \cong W(k,Z;D)$. By this

we mean we are free to choose localizers embedding the residue field
$D/P$ into $E/\Delta^{-1}(D/Z)$ which specify the isomorphism. <u>However</u>, in
order that the isomorphisms determined by our particular choice of
localizers be applicable to computing $\partial(S)$, the particular choice of
localizers given must make our previous diagram commute.  ie.

$$
\begin{array}{ccc}
H(E/\Delta^{-1}(D/Z)) & \to & \oplus\, H(D/P) \\
\downarrow t & & \downarrow tr \\
W(k,Q/Z;D) & \xrightarrow{\;g\;} & \underset{P=\overline{P}}{\oplus}\, W(k,F_p;D/P)
\end{array}
$$

There are many ways to give the isomorphism $H(E/\Delta^{-1}(D/Z)) \simeq \oplus\, H(D/P)$.
As we see from the preceding, there is one canonical isomorphism, namely
$tr^{-1} \circ g \circ t$. We shall see in the next section that there is also a
convenient way to give this isomorphism in order to read $\partial(D,P)$.

Thus, we have the following plan.

(1)  First we study the case of a maximal order.  We shall
use a convenient choice of localizers to compute $\partial(D,P)$.  This enables
us to compute $\partial(D,P)$ and the cokernel of $\partial(D)$.

(2)  In Chapter 7 we discuss non-maximal orders.  We must
discuss three key parts in this regard.

(a)  We discuss the canonical localizers which make the finite
field trace and number field trace induced isomorphisms
agree.

(b)  We compute trace induced map $tr: H(D/P) \to H(S/P \cap S)$
for finite fields.

(c)  We describe the set $T(M)$.

We shall next be computing the group $W(k,Z;D) \simeq H(\Delta^{-1}(D/Z))$ by using the exact sequence from Proposition 2.3.

$$0 \to H(\Delta^{-1}(D/Z)) \to H(E) \xrightarrow{\partial(D)} H(E/\Delta^{-1}(D/Z))$$

We recall that $H(E)$ for $E$ an algebraic number field, and $H(E/\Delta^{-1}) \simeq \oplus H(D/P)$ are known. (Here we abbreviate $\Delta^{-1}(D/Z)$ by $\Delta^{-1}$).

In analyzing the boundary map $\partial(D)$, we reduce to the local case by using:

<u>Proposition 2.7</u> <u>Localization at a - invariant, maximal ideal</u> $P$ <u>in</u> D <u>induces the commutative diagram:</u>

$$\begin{array}{ccccccc} 0 & \to & H(\Delta^{-1}(D/Z)) & \to & H(E) & \xrightarrow{\partial(D)} & H(E/\Delta^{-1}) \simeq \bigoplus_{P=\overline{P}} H(D/P) \\ & & \downarrow & & \downarrow & & \downarrow \qquad\qquad \downarrow q(P) \\ 0 & \to & H(\Delta^{-1}(P)) & \to & H(E) & \xrightarrow{\partial(D,P)} & H(E/\Delta^{-1}(P)) \simeq H(D/P) \end{array}$$

<u>This diagram also commutes for an order</u> S.

<u>Proof</u>: $\partial(D,P) = q(P) \circ \partial(D)$ by definition. What we show here is that these maps are induced by localization.

As long as S is an order in D, maximal or not, S is a finitely generated Z-algebra, and hence Noetherian by [A,M 81]. Thus by [B-2 20], every finitely generated S-module M is finitely presented. Hence [B-2 76] the adjoint isomorphism $M \to \mathrm{Hom}(M,E/\Delta^{-1})$ localizes to $M(M) \to \mathrm{Hom}_{S(M)}(M(M), (E/\Delta^{-1})(M))$, where $M$ is a maximal ideal in S. Thus localization preserves the non-singularity of the forms, and indeed defines a map.

In order to identify $(E/\Delta^{-1})(P)$ with $E/\Delta^{-1}(P)$, use the exact sequence

$$0 \to \Delta^{-1} \to E \to E/\Delta^{-1} \to 0,$$

and [A,M 39]. □

Comment: This re-emphasizes the remarks before Theorem IV 1.9.

Corollary 2.8 $H(\Delta^{-1}(D/Z)) = \bigcap_{P=\overline{P}} H(\Delta^{-1}(D/Z)(P))$

Proof:
$$\begin{aligned} H(\Delta^{-1}) &= \ker \partial(D) \\ &= \bigcap_{P=\overline{P}} \ker q(P) \circ \partial(D) \\ &= \bigcap_{P=\overline{P}} \ker \partial(D,P) \\ &= \bigcap_{P=\overline{P}} H(\Delta^{-1}(P)) \end{aligned}$$
□

The computation of $\partial(D,P)$ will be made in Section 3.

## 3. Computing the local boundary $\partial(D,P)$ for the maximal order.

We consider the general case $\partial(D): H_u(E) \to H_u(E/I)$ where $I = \overline{I}$ is a - invariant fractional ideal. Of course, we have in mind $I = \Delta^{-1}(D/Z)$.

Following Proposition 2.7, we wish to compute the localization, $\partial(D,P)$ of $\partial(D)$ at a - invariant maximal ideal $P = \overline{P}$ in D. Since D will be fixed throughout this section, we simplify our notation of $\partial(D,P)$ to $\partial(P)$.

From 2.7, $(E/I)(P) = E/I(P)$. We now consider $E/I(P)$. (See IV 3.8). We embed the residue field $D/P$ into $E/I(P)$ as follows. Let $\rho \in E$ satisfy $v_P(\rho) = v_P(I) - 1$. Then define

$$f: D/P \rightarrow E/I(P)$$

by

$$r + P \rightarrow \rho r + I(P).$$

f is $1 - 1$ since $\rho r \in I(P)$ implies

$$v_P(\rho r) \geq v_P(I(P)) = v_P(I)$$

so

$$v_P(I) - 1 + v_P(r) \geq v_P(I)$$

$$v_P(r) \geq 1$$

Hence $r \in P$.

For the case when $S$ is an order in $E$, with $I = \Delta^{-1}(S/Z)$, we claim that there is a commutative diagram

$$\begin{array}{ccccc} 0 & \rightarrow & S/M & \xrightarrow{\tilde{w}} & E/I(M) \\ & & \downarrow tr & & \downarrow t \\ 0 & \rightarrow & F_p & \xrightarrow{w} & Q/Z(p). \end{array}$$

Here tr denotes trace on the finite field level. t is induced by trace of $E/Q$. w is given by the canonical choice of uniformizer in $Q/Z(p)$ annihilated by p, namely $w: 1 \rightarrow (\frac{1}{p})$.

In order to see that $\tilde{w}$ exists we proceed as follows. Let A be a finitely generated S-module. Then S is the image of a free

S-module, $F_1$, and we have the exact sequence $0 \to \ker f \to F_1 \xrightarrow{f} A \to 0$. Of course $F_1/\ker f \approx A$. This leads to the diagram below, with $F_2 = \ker f$.

$$\begin{array}{ccc} 0 & & 0 \\ \downarrow & & \downarrow \\ F_2 & \xrightarrow{h_2} & Z \\ \downarrow & & \downarrow \\ F_1 & \xrightarrow{h_1} & Q \\ \downarrow & & \downarrow \\ A & \xrightarrow{h} & Q/Z \end{array}$$

Given $h \in \mathrm{Hom}_Z(A,Q/Z)$, $h$ lifts to $h_1: F_1 \to Q$ since $F_1$ is projective. By commutativity, $h_1|_{F_2} = h_2 \in \mathrm{Hom}_Z(F_2,Z)$.

For finitely generated projective S-modules A, we assert that there is a trace $E/Q = t$ induced isomorphism:

$$\mathrm{Hom}_S(A,E) \xrightarrow{\hat{t}} \mathrm{Hom}_Z(A,Q). \qquad \hat{t}(g) = t \circ g$$

$\hat{t}$ is clearly onto since A is S-projective. In order to see $\hat{t}$ is 1 - 1, suppose $g \in \mathrm{Hom}_S(A,E)$. Let $a \in A$ satisfy $g(a) \neq 0$. Then clearly there exists $e \in E$ with $t(eg(a)) \neq 0$. However S is an order in E so we can write $m \cdot e \in S$ for some $m \in Z$. Thus, since t is Z-linear,

$$m \cdot t(eg(a)) = t(me\, g(a)) = t(g(mea)) \neq 0.$$

It follows that $t \circ g \neq 0$, and $\hat{t}$ is an isomorphism.

Hence $h_1 \in \mathrm{Hom}_Z(F_1, Q)$ may be written uniquely as $t \circ k_1$ where $k_1 \in \mathrm{Hom}_S(F_1, E)$. Further, since $t \circ k_1|_{F_2} = h_2$, we observe that

$$k_1|_{F_2} = k_2 \in \mathrm{Hom}_S(F_2, \Delta^{-1}(S/Z)).$$

Thus $k_1$ induces an S-module homomorphism $k \in \mathrm{Hom}_S(F_1/F_2, E/\Delta^{-1}(S/Z))$. Clearly $t \circ k = h$.

We claim that this $k$ is unique. For suppose $t \circ j = t \circ k = h$. Then $t \circ (j - k) = 0$ in $Q/Z$. Now consider the diagram

$$\begin{array}{ccc} F_1 & \xrightarrow{\widehat{(j - k)}} & E \\ \downarrow & & \downarrow \\ F_1/F_2 & \xrightarrow{(j - k)} & E/\Lambda^{-1} \end{array}$$

$\widehat{(j - k)}$ exists since $F_1$ is S-projective.

However, we also have the commutative diagram:

$$\begin{array}{ccc} E & \xrightarrow{t} & Q \\ \downarrow & & \downarrow \\ E/\Delta^{-1} & \xrightarrow{t} & Q/Z \end{array}$$

Thus $t \circ \widehat{(j - k)} \subset Z$, from which it follows that we have $\mathrm{im}\widehat{(j - k)} \subset \Delta^{-1}(S/Z)$. Hence $j - k \equiv 0$ as maps in $\mathrm{Hom}_S(F_1, E/\Delta^{-1}(S/Z))$ and $j = k$ is unique.

We apply this to the finitely generated S-module $S/M$, where $h: S/M \to Q/Z(p)$ is the $Z(p)$-module homomorphism $h = w \circ tr$.

By the above, there exists a unique $k = \tilde{w}$ with $t \circ \tilde{w} = w \circ tr$ as claimed.

$\tilde{w}$ is evidently determined by where $\overline{1} \in S/M$ is taken. Hence $\tilde{w}$ is in fact determined by $\rho_M$ for a suitable choice of localizer $\rho_M$.

However, in our computation of $\partial(P)$, we shall find it convenient to specify the localizers $\rho_P$ in a different manner. The manner in which we pick these is dictated by our desire to have the boundary computation read by Hilbert symbols.

<u>Theorem 3.1</u> Let $[M,B] \in H_u(E)$. <u>We diagonalize</u> B <u>as</u> $B = \langle a_1x_1\rangle \oplus \ldots \oplus \langle a_nx_1\rangle$ <u>where</u> $a_i \in F/NE$, <u>and</u> $x_1\overline{x}_1^{-1} = u$ <u>is chosen in a prescribed manner to be described in the proof. Having fixed our choice of</u> $x_1$, <u>there is a choice of localizers</u> $\rho_P$ <u>so that the following holds:</u>

I. <u>If</u> $P$ <u>is over inert, so that</u>

$$\partial(P): H_u(E) \to H_u(E/I(P)) \stackrel{\cong}{\to} H(D/P) \cong \{0,1\}$$

<u>we have the following formulas.</u>

(a) <u>If</u> $B = \langle ax_1\rangle$ <u>is of rank</u> 1,

$$(a,\sigma)_p = (-1)^{\partial(P)\langle ax_1\rangle + v_P(I)}.$$

(b) <u>If</u> B <u>has even rank and discriminant</u> d <u>relative to</u> $x_1$, <u>then:</u> $(d,\sigma)_p = (-1)^{\partial(P)(B)}$

(c) If there are no ramified primes, and $x_1$ has odd valuation $v_P(x_1)$ at an even number of primes formulas (a) and (b) above hold.

(d) If there are no ramified primes, and $x_1$ has odd valuation at an odd number of primes, formulas (a) and (b) are valid at all inert primes except one specified prime $P_1$ over $p_1$. At $P_1$, $v_{P_1}(ax_1) = v_P(a) + 1$ and we then have

$$\text{(a)}' \qquad (a,\sigma)_{P_1} = (-1)^{\partial(P)\langle ax_1\rangle + v_P(I) + 1}$$

(b) For B of even rank, formula (b) still holds.

II. Ramified primes are divided into two classes:

(a) $cl(P) = 0$ if $v_P(I) \equiv v_P(x_1) \pmod 2$

(b) $cl(P) = 1$ if $v_P(I) \equiv v_P(x_1) + 1 \pmod 2$

$\partial(P) = 0$ at ramified primes of class 0. $\partial(P)$ preserves rank at ramified primes of class 1.

Note: This determines $\partial(P)$ at dyadic ramified primes.

Further, under the choice of localizers $\rho_P$ made, $H_u(E/I(P)) \approx W(D/P)$ reads $\partial(P)$ as follows: $\partial(P)(\langle ax_1\rangle)$ is a non-square if and only if $(a,\sigma)_P = -1$. Further, if B has even rank, and discriminant d, then $\partial(P)(B)$ has a non-trivial discriminant if and only if $(d,\sigma)_P = -1$.

Proof: We begin by considering u, with $u\bar{u} = 1$. By Hilbert's Theorem 90, there exists $x \in E^*$ with $x\bar{x}^{-1} = u$. Our first task as described in the theorem is to rechoose x appropriately.

Thus, we consider $v_P(x)$, for $P$ over inert. If at least one prime ramifies, finite or infinite, we can find $y \in F^*$ with $(y,\sigma)_P = (-1)^{v_P(x)}$ for all $P = P \cap F$ which are inert, by realization of Hilbert symbols. If there are no ramified primes, finite or infinite, there are two possibilities.

(1)  If $v_P(x)$ is odd at an even number of inert primes, by Realization it is still possible to choose $y \in F^*$ with $(y,\sigma)_P = (-1)^{v_P(x)} = (-1)^{v_P(y)}$ at all inerts.

(2)  If $v_P(x)$ is odd at an odd number of inert primes, we may find $y \in F^*$ with $(y,\sigma)_P = (-1)^{v_P(x)}$ at all inert primes except one specific inert prime, say $P_1$, at which

$$(y,\sigma)_{P_1} = (-1)^{v_{P_1}(x) + 1} \qquad P_1 \cap F = P_1.$$

We now rechoose $x_1 = xy$. We still have $x_1 \bar{x}^{-1} = u$ since $y \in F^*$ has $y\bar{y}^{-1} = yy^{-1} = 1$. Note, however, that $x_1$ now has even valuation at all inert primes, with at most one exception as described above.

We next describe how to choose the localizers $\rho_P$.

First, at $P$ inert, we choose $\rho_P = x_1 w$, where $w \in F^*$ satisfies $v_P(w) = v_P(I) - v_P(x) - 1$, so $v_P(\rho_P) = v_P(I) - 1$.

This is possible at inert $P$, since a local uniformizer for $P \cap F$ is also a local uniformizer for $P$.

In order to describe the ramified primes $P$ we begin as follows. We note that the $-$ involution makes the local units in $O_E(P)$, denoted $O_E(P)^*$ into a $C_2$-module.

<u>Lemma 3.2</u> <u>If</u> $P = \overline{P}$ <u>is over inert, then</u> $H^1(C_2;O_E(P)^*) = 1$. <u>If</u> $P = \overline{P}$ <u>is over ramified</u>, $H^1(C_2;O_E(P)^*) = C_2$.

<u>Proof</u>: Recall that $H^1(C_2;O_E(P)^*) = \{x \in O_E(P)^*: x\overline{x} = 1\}$ modulo $\{v/\overline{v}: v \in O_E(P)^*\}$.

If $x$ is a local unit of norm 1, then by Hilbert 90, there exists $z \in E^*$ with $z\overline{z}^{-1} = x$.

Write $z = \pi^{v_P(z)}v$, where $\pi$ is a local uniformizer for $P$, and $v \in O_E(P)^*$. If $P$ is over inert, we may choose $\pi \in F^*$, so that $\pi = \overline{\pi}$. Thus $z\overline{z}^{-1} = v\overline{v}^{-1} = x$, and $H^1$ is trivial in this case.

If $P$ is over ramified, $(\pi\overline{\pi}^{-1})$ is a local unit, and $z\overline{z}^{-1} = (\pi\overline{\pi}^{-1})^{v_P(z)}v\overline{v}^{-1} = x$. Thus $H^1$ is generated by the class of $\pi\overline{\pi}^{-1}$, $cl(\pi\overline{\pi}^{-1})$. Of course $(\pi\overline{\pi}^{-1})^2$ is trivial in $H^1$, since $(\pi\overline{\pi}^{-1})^2 = (\pi\overline{\pi}^{-1})(\overline{\pi}\pi^{-1})^{-1}$, a quotient of local units. Thus, to complete the proof we need only show that $cl(\pi\overline{\pi}^{-1})$ is non-trivial in $H^1$.

Suppose to the contrary that there is a local unit $v$ with $v\overline{v}^{-1} = \pi\overline{\pi}^{-1}$. Then $\pi v^{-1} = \overline{\pi}\overline{v}^{-1}$, so that $\pi v^{-1}$ is a local uniformizer of $O_E(P)$ which lies in $F$. This is impossible as we are in the ramified case. Hence, $cl(\pi\overline{\pi}^{-1})$ is non-trivial as claimed.

We now consider $cl(u) \in H^1(C_2;O_E(P)^*)$ where $P = \overline{P}$ is over ramified. By Lemma 3.2, $H^1(C_2;O_E(P)^*) \neq 0$ and we may write

$$cl(u) = cl(\pi\overline{\pi}^{-1})^{v_P(I) - \varepsilon} \qquad \text{where } \varepsilon = 0 \text{ or } 1.$$

<u>Definition 3.3</u> <u>If</u> $P = \overline{P}$ <u>is over ramified</u>, $P$ <u>is of class</u> 0 <u>if</u> $\varepsilon = 0$. $P$ <u>is of class</u> 1 <u>if</u> $\varepsilon = 1$.

<u>We observe the following:</u>

<u>Lemma 3.4</u> <u>We may rephrase class as follows:</u>

(a) $cl(P) = 0$ <u>if and only if</u> $v_P(I) \equiv v_P(x_1) \pmod 2$

(b) $cl(P) = 1$ <u>if and only if</u> $v_P(I) \equiv v_P(x_1) + 1 \pmod 2$

<u>Proof</u>: Here $x_1\bar{x}_1^{-1} = u$. Write $x_1 = \pi^i w$, $w \in O_E(P)^*$. Then

$$cl(u) = cl(x_1\bar{x}_1^{-1}) = cl(\pi\bar{\pi}^{-1})^i$$

However, $cl(\pi\bar{\pi}^{-1})$ generates $H^1(C_2;O_E(P)^*) \neq 0$, from which the result follows. □

When $P$ is tamely ramified, we choose $\pi = -\bar{\pi}$, a skew - uniformizer. When the ramification is wild, any uniformizer will do. Note that $\pi\bar{\pi}$ is a uniformizer for $P \cap F = P$, since $P$ is over ramified. Also, $v_P(\pi\bar{\pi}) = 2$. We now choose $\rho_P$ at ramified primes as:

$\rho_P = x_1(\pi\bar{\pi})^t\pi$ with $t$ suitably chosen so that

$v_P(\rho_P) = v_P(I) - 1$ if $cl(P) = 0$.

$\rho_P = x_1(\pi\bar{\pi})^t$ with $t$ suitably chosen so that

$v_P(\rho_P) = v_P(I) - 1$ if $cl(P) = 1$.

With these choices of localizers made, we now identify the image groups of $\partial(P)$, $H_u(E/I(P))$. Let $[V,B] \in H_u(E/I(P))$. By assuming that $(V,B)$ is anisotropic, it follows that the annihilator of the finitely generated $O_E(P)$ - module $V(P)$ is the maximal ideal $m(P)$

in $O_E(P)$. Thus $V$ is an $O_E(P)/m(P)$ - module. This is equivalently phrased by saying $V$ is a vector space over the residue field $O_E(P)/m(P) = D/P$.

Let $x,y \in V$. Suppose $B(x,y) = [a] \in E/I(P)$. Letting $\pi$ be a uniformizer for $P$ as above, $\pi x = 0$ since $V$ is an $O_E(P)/m(P)$ - module. Thus $\pi[a] \in I(P)$.

Let $a_1$ be a lift of $a$ to $E$. Since $\pi a_1 \in I(P)$, it follows that $v_P(a_1) \geq v_P(I) - 1$. Also $B$ is u Hermitian, so that $[a] = u[\bar{a}]$ in $E/I(P)$. We may thus write $a_1 - u\bar{a}_1 \in I(P)$.

We now consider the $D/P$ - valued form on $V$ given by $B^1 = \rho_P^{-1} \cdot B$, where the choice of $\rho_P$ has been previously specified.

(1) $P$ inert. With $a_1 - u\bar{a}_1 = i \in I(P)$, we show $B^1$ is +1 Hermitian. Here $\rho_P = x_1 w$ where $w \in F^*$.

$$
\begin{aligned}
\overline{a_1 \rho_P^{-1}} &= \left(\frac{a_1 - i}{u}\right)\bar{x}_1^{-1}\,\bar{w}^{-1} \\
&= \left(\frac{a_1 - i}{\not u}\right)\left(\frac{\not u}{x_1}\right)w^{-1} \\
&= \frac{a_1 w^{-1}}{x_1} - i\frac{w^{-1}}{x_1} = a_1\rho_P^{-1} - i\rho_P^{-1} \\
&= a_1\rho_P^{-1} \text{ in } D/P.
\end{aligned}
$$

This last follows because

$$v_P(i\rho_P^{-1}) = v_P(i) - v_P(I) + 1 \geq 1, \text{ so that } i\rho_P^{-1} \in P.$$

This shows there is an isomorphism between

$$H_u(E/I(P)) \quad \text{and} \quad H_{+1}(O_E(P)/m(P))$$

given by scaling with $\rho_P$. Since $P$ is inert, the involution induced on $O_E(P)/m(P)$ is non-trivial, and we have true $+1$ Hermitian.

(2) The tamely ramified case.
As before, we have the form $B^1 = \rho_P^{-1}B$.

(a) $cl(P) = 0$, so $\rho_P = x_1(\pi\bar{\pi})^t\pi$ where $\pi = -\bar{\pi}$

We now compute as before:

$$\bar{a}_1\bar{\rho}_P^{-1} = \left(\frac{a_1 - i}{\not{2}}\right)(\bar{x}_1^{-1})(\pi\bar{\pi})^{-t}(\bar{\pi})^{-1}$$

$$= \left(\frac{a_1 - i}{\not{2}}\right)\left(\frac{\not{2}}{x_1}\right)(\pi\bar{\pi})^{-t}(-\pi)^{-1}$$

$$= (a_1 - i)(-\rho_P^{-1}) = -a_1\rho_P^{-1} \quad \text{in} \quad O_E(P)/m(P).$$

Since $P$ is ramified, we obtain this time an isomorphism between $H_u(E/I(P)) \simeq W^{-1}(D/P) = 0$.

(b) $cl(P) = +1$, so $\rho_P = x_1(\pi\bar{\pi})^t$. The same computation shows $\bar{a}_1\bar{\rho}_P^{-1} = a_1\rho_P^{-1}$ in $D/P$, and we have $H_u(E/I(P)) \simeq W^{+1}(D/P)$.

(3) The case for wild ramification follows as above. In either case, $H_u(E/I(P)) \simeq W(D/P)$.

With these preliminaries, we are ready to compute $\partial(P)$. To begin with, consider a 1-dimensional form in $H_u(E)$. By II 4.15 we may write this as $\langle ax_1\rangle$, for $x_1$ fixed as described, and $a$ uniquely determined in $F/N_{E/F}E$.

I. We first compute $\partial(P)$ for $P$ over inert. We begin by considering $\partial(P)(\langle ax_1\rangle)$. Observe the Witt equivalence

$$\langle ax_1\rangle \sim \langle ax_1(\pi\bar{\pi})^t\rangle,$$

for $\pi$ a uniformizer for $P$. It follows that without loss of generality, we may assume either:

(a) $v_P(ax_1) = v_P(I) - 2$

or

(b) $v_P(ax_1) = v_P(I) - 1$.

This is done by rechoosing $a$ as $a(\pi\bar{\pi})^t$. Here, recall we may choose $\pi \in P \cap F = P$ since this is the inert case. In any case, $v_P(\pi\bar{\pi}) = 2$ and $\pi\bar{\pi}$ is not a uniformizer for $P$. Thus case (a) or (b) only depends on $v_P(a)$ compared to $v_P(I)$, since by choice $v_P(x_1) \equiv 0 \pmod 2$ with at most one exceptional prime $P_1$.

Now consider the lattice $L = P$. Since $\langle ax_1\rangle$ has $v_P(ax_1) = v_P(I) - 1$ or $v_P(I) - 2$, $\langle ax_1\rangle|_P$ is I-valued.

We consider the dual lattice:

$$\begin{aligned} L^{\#} &= \{x \in E: B(x,L) \subset I\} \\ &= \{x \in E: xax_1\bar{P} \subset I\} \\ &= \{x \in E: x \in I\bar{P}^{-1}(ax_1)^{-1}\} \\ &= \begin{cases} P & \text{in case (a)} \\ O_E(P) & \text{in case (b)} \end{cases} \end{aligned}$$

Thus viewed, in case (a), we clearly get $\partial(P)\langle ax_1\rangle = 0$.

In case (b), $\partial(P)\langle ax_1\rangle = \langle O_E(P)/m(P), B^1\rangle$, where $B^1$ is defined on the torsion $O_E(P)$ - module $L^{\#}/L = O_E(P)/m(P)$ with values in $E/I(P)$ by $\langle ax_1\rangle$, with $v_P(ax_1) = v_P(I) - 1$. As we have mentioned, we then identify this with the $D/P$ - valued form on $D/P$ given by $\langle ax_1\rho_P^{-1}\rangle$, where $\rho_P = x_1 w$, ie.

$$\langle ax_1\rho_P^{-1}\rangle = \langle ax_1 x_1^{-1} w\rangle = \langle aw\rangle \qquad w \in F^*.$$

Again, $P$ is inert so that $[O_E(P)/m(P): O_F(P)/m(P)] = 2$ and the induced involution on the finite field $O_E(P)/m(P)$ is non-trivial. Hermitian of a finite field is determined by rank modulo 2. Thus, we have completed our computation of the local boundary on a 1-dimensional form when $P$ is inert.

Identifying $H_u(E/I(P)) \cong H_{+1}(D/P) \cong F_2 = \{0,1\}$, we may summarize this as:

$$(-1)^{\partial(P)\langle ax_1\rangle} = (-1)^{v_P(ax_1) - v_P(I)}$$

or

$$\begin{aligned}(-1)^{\partial(P)\langle ax_1\rangle + v_P(I)} &= (-1)^{v_P(ax_1)}\\ &= (-1)^{v_P(a)}\\ &= (a,\sigma)_P.\end{aligned}$$

Continuing in the inert case, let $B$ be a form of even rank. Since $E$ is a field, we may diagonalize $B$ as before,

$$B = \langle a_1x_1\rangle \oplus \langle a_2x_1\rangle \oplus \dots \oplus \langle a_nx_1\rangle, \text{ where } a_i \in F/NE.$$

As in II 4.15, we define the discriminant of $B$ to be $(-1)^{\frac{n(n-1)}{2}} \Pi a_i$. Note that this depends on the fixed choice of $x_1$. Adding a hyperbolic form $\langle x_1\rangle \oplus \langle -x_1\rangle$ if necessary we may, without loss of generality, take $n$ to be a multiple of 4. This does not effect $\partial(P)$ or $d$, but it does enable us to write $d = \Pi a_i$. We now use that $\partial(P)$ is additive, to compute:

$$\begin{aligned}
(d,\sigma)_P &= (a_1 \dots a_n,\sigma)_P \\
&= \prod_i (a_i,\sigma)_P \\
&= \prod_i (-1)^{\partial(P)\langle a_i x_i\rangle} + v_P(I) \\
&= \prod_i (-1)^{\partial(P)\langle a_i x_i\rangle} \\
&= (-1)^{\partial(P)[B]}.
\end{aligned}$$

This completes the inert case.

II. <u>The ramified case.</u>

As in the inert case, we begin by considering a 1-dimensional form $\langle ax_1\rangle$ in $H_u(E)$. Here $a$ is unique in $F/NE$, $a \in F$, so $v_P(a) \equiv 0 \pmod 2$, since $P$ is ramified. Scaling $a$ by the norm $(\pi\bar{\pi})$ from $E$, we may assume $v_P(a) = 0$.

Note: This does not affect $(a,\sigma)_P$, nor $\partial(P)\langle ax_1\rangle$. We now scale the resulting form $\langle ax_1\rangle$ and obtain the Witt-equivalent form $\langle ax_1(\pi\bar{\pi})^t\rangle$, with

$$v_P(ax_1(\pi\bar{\pi})^t) = v_P(x_1(\pi\bar{\pi})^t) = \begin{cases} v_P(I) - 2 \\ v_P(I) - 1 \end{cases}$$

depending on $cl(P)$.

As in the inert case, we let $L = P$, and compute

$$L^{\#} = P \quad \text{if} \quad cl(P) = 0$$
$$= O_E(P) \quad \text{if} \quad cl(P) = 1$$

Thus, if $cl(P) = 0$, $\partial(P) = 0$. If $cl(P) = 1$, we obtain the $E/I(P)$ - valued form $<O_E(P)/m(P), B'>$, where $B' = <ax_1(\pi\bar{\pi})^t>$. We identify this with the $O_E(P)/m(P)$ - valued form on $O_E(P)/m(P)$ given by viewing $a$ in $O_E(P)/m(P)$: $<ax_1(\pi\bar{\pi})^t\rho_P^{-1}> = <a>$. Note that this is a Witt inner product since $P$ ramifies, so that

$$[O_E(P)/m(P): O_F(P)/m(P)] = 1,$$

and the induced involution on the finite field $O_E(P)/m(P)$ is trivial.

(1) If the characteristic of $O_E(P)/m(P) = 2$, rank is the only invariant, and we are done.

(2) If the characteristic of $O_E(P)/m(P) \neq 2$, we must determine if $a$ is a square in the residue field.

By II 2.4, $a$ is a square in $O_E(P)/m(P)$ if and only if $(a,\sigma)_P = +1$.

We continue by letting $B$ be a form of even rank. As before, we diagonalize $B$, $B = <a_1x_1> \oplus \ldots \oplus <a_nx_1>$. Again, without loss of generality, $n \equiv 0 \pmod 4$. By additivity of the boundary,

$\partial(P)(B) = \langle a_1 \rangle \oplus \ldots \oplus \langle a_n \rangle$, which has discriminant $\prod_{i=1}^{n} a_i$, since $n \equiv 0 \quad (4)$. Again, $\prod_{i=1}^{n} a_i$ is a square in $D/P$ if and only if $\prod_{i=1}^{n} a_i = d$ is a local norm, if and only if $(d,\sigma)_P = +1$. This completes the computation of $\partial(P)$. □

4. Computing the cokernel of $\partial(D)$.

In this section, we use the computation of the local boundary, $\partial(P)$, to compute

$$\partial\colon\ H_u(E) \to H_u(E/I).$$

We also show how to compute $H_u(I)$, where $I = \Delta^{-1}(D/Z)$. Of course, $H(\Delta^{-1}(D/Z) \cong W(k,Z;D)$, and $H_u(\Delta^{-1}(D/Z)) = A(Z;D)$. Further, the computation of the boundary on the Hermitian level will subsequently be used in the computation of the global boundary $\partial\colon W(k,Q) \to W(k,Q/Z)$ in Chapter VIII.

In order to describe the boundary homomorphism, the complicated case is described in the next Lemma.

Lemma 4.1 Suppose $E$ has involution $-$, fixed field $F$ as usual. Suppose $E/F$ has no signatures, no dyadic ramified primes, and all ramified primes are of class 1. Then we may write the collection of ramified primes as $P_1, \ldots, P_{2t}, P_1', \ldots, P_r'$ where the $P_i$, $i = 1 \ldots 2t$ have residue fields $O_E(P_i)/m(P_i) = F_q$ with $q \equiv 3 \quad (4)$ and the $P_i'$, $i = 1, \ldots r$ have residue fields $F_q$ with $q \equiv 1 \quad (4)$ (assuming $2t + r \neq 0$).

Proof: We wish to show that the number of ramified primes whose residue fields $F_q$ have $q \equiv 3 \ (4)$ is even. Let $P_i$ be such a prime. Then $-1$ is not a square in each $O_E(P_i)/m(P_i) = F_q$ [Lm 43]. However at ramified primes the square class in the residue field determines the Hilbert symbol. Thus $(-1,\sigma)_{P_i} = -1$ at each such $P_i = P_i \cap F$ over ramified. Notice that at the other ramified primes, $P_i'$, whose residue fields are $O_E(P_i')/m(P_i') = F_q$ with $q \equiv 1 \pmod 4$, $-1$ is a square in the residue field, so that $(-1,\sigma)_{P_i'} = +1$. Also, $-1$ is a local unit at inerts, so that

$$(-1,\sigma)_P = (-1)^{v_P(-1)} = +1 \text{ at all inerts.}$$

We now apply Hilbert reciprocity, $\prod_i(-1,\sigma)_{P_i} = +1$. This shows that the number of primes with residue fields $F_q$ with $q \equiv 3 \ (4)$ elements must be even, since $(-1,\sigma)_{P_i} = -1$ only at those primes.

Next we form the group $G$ given by

$$G = Z^* \ltimes (F_2 \times \ldots \times F_2)^{2t} \times (F_2 \times \ldots \times F_2)^r,$$

where $2^t$ = number of ramified primes at which $(-1,\sigma)_P = -1$, $r$ = number of ramified primes at which $(-1,\sigma)_P = +1$, as given by Lemma 4.1. We write $Z^* = \{1,-1\}$, $F_2 = \{0,1\}$. $Z^*$ is designed to keep track of the discriminant and reciprocity, $F_2$ will take care of ranks.

On $G$ we define a multiplication as follows:

$(c,a_1, \ldots a_{2t}, b_1, \ldots b_r) \dot{x} (c', a_1', \ldots a_{2t}', b_1', \ldots b_r')$

$= ((-1)^{a_1a_1' \ldots + a_{2t}a_{2t}'}cc', a_1 + a_1', \ldots a_{2t} + a_{2t}', b_1 + b_1', \ldots$

$, b_r + b_r')$.

Note that $(1,0, \ldots, 0)$ is the identity in $G$, and that the order of every element divides $4$.

The purpose of this group $G$ is to describe the cokernel of $\partial\colon H_u(E) \to H_u(E/I)$ in the special case that there are no signatures, no dyadic ramified primes, and there are ramified primes, all of which are of class 1 in the extension $E/F$. We assume now that we are in this case.

To begin with, recall that

$$H_u(E/I) \simeq \bigoplus_{P = \bar{P} \text{ inert}} H(D/P) \quad \bigoplus_{P = \bar{P} \text{ ramified}} W(D/P).$$

Thus any element in $H_u(E/I)$ can be expressed as a direct sum of elements $\oplus [M_iB_i]$, where $[M_iB_i]$ is either in $H(D/P)$ or $W(D/P)$. We define a map $h\colon H_u(E/I) \to G$ by:

Let $[M_i,B_i] \in H(D/P)$. This Hermitian element depends only on the rank modulo 2 of $M_i$ over $D/P$. We define:

$$h([M_i,B_i]) = ((-1)^{\text{rank } M_i}, 0, 0, \ldots 0) \in G$$

for these Hermitian summands.

Continuing, suppose $\bigoplus_P [M_i,B_i] \in \bigoplus_{P \text{ ramified}} W(D/P)$

Let $d_i$ = discriminant of $B_i$

$c = \prod_{P_i \text{ ramified}} (d_i,\sigma)_{P_i}$

$a_i$ = local rank of $M_i$ over $D/P_i$ $\quad i = 1, \ldots 2t$

$b_i$ = local rank of $M_i$ over $D/P_i'$ $\quad i = 1, \ldots r$

Define $h\colon \bigoplus_P [M_i,B_i] \to (c, a_1, \ldots a_{2t}, b_1, \ldots b_r)$. Clearly, by additivity, this defines $h$ as a map $h\colon H_u(E/I) \to G$.

<u>Lemma 4.2</u> $h$ <u>is</u> <u>a</u> <u>homomorphism</u>.

<u>Proof</u>: This follows using the product formula for discriminants, II 4.10. The local discriminants satisfy

$$\text{dis}\,([M_1,B_1] \oplus [M_2,B_2]) = (-1)^{\text{rank } M_1 \text{ rank } M_2}\,\text{dis } B_1 \cdot \text{dis } B_2$$

We now use the following abbreviations:

$$\text{dis } B_1 = d_1 \quad \text{dis } B_2 = d_1' \quad \text{rank } M_1 = a_1 \quad \text{rank } M_2 = a_1'.$$

At each of the first $2t$ ramified primes,

$$\begin{aligned}((-1)^{a_1a_1'}d_1d_1', \sigma)_P &= (-1)^{a_1a_1'}(d_1d_1',\sigma)_P\\ &= (-1)^{a_1a_1'}(d_1,\sigma)_P(d_1',\sigma)_P\end{aligned}$$

since $(-1)$ is not a square in $D/P$. At each of the next $r$ ramified primes,

$$((-1)^{b_1 b_1'} d_1 d_1', \sigma)_P = (d_1 d_1', \sigma)_P = (d_1, \sigma)_P (d_1', \sigma)_P$$

since $(-1)$ is a square in $D/P$.

From these formulas, it clearly follows that $h$ is a homomorphism. □

<u>Lemma 4.3</u> <u>The image of</u> $\partial$ <u>in</u> $H_u(E/I)$ <u>is mapped under</u> $h$ <u>to the subgroup</u> $W_2$ <u>of</u> $G$ <u>whose elements are</u> $(1,0,\ldots 0)$ <u>and</u> $(\varepsilon,1,1,\ldots 1)$ <u>where</u> $\varepsilon$ <u>is given by</u> $\varepsilon = \prod_{P \text{ inert}} (-1)^{v_P(I)}$. (<u>same hypotheses as</u> 4.1)

<u>Proof</u>: Let $\bigoplus_{P \text{ inert}} [M_i, B_i] \bigoplus_{P \text{ ramified}} [M_i, B_i]$ be in the image of $\partial$, say $\partial[M,B]$ equals this element in $H_u(E/I)$. In other words $\partial(P_i)[M,B] = [M_i, B_i]$, where $[M_i, B_i] \in H(D/P_i)$ or $W(D/P_i)$ depending on whether $P_i$ is over inert or ramified.

Case (1) $M$ has even rank, and discriminant $d$. By Theorem 3.1, $\partial(P)$ is read by the Hilbert symbol, $(d,\sigma)_P = (-1)^{\partial(P)[M,B]}$. Rank is preserved at class 1 ramified primes. Hence by Hilbert reciprocity,

$$h \circ \partial\,[M,B] = (1, 0, \ldots 0) \in G.$$

Case (2) $M$ has odd rank, and discriminant $d$. At inert primes:

$$(d,\sigma)_P = (-1)^{\partial(P)[M,B] + v_P(I)}.$$

At ramified primes, the square class of the discriminant of $\partial(P)[M,B]$ is determined by the Hilbert symbol $(d,\sigma)_P$. Rank mod 2 is preserved, since by hypothosis all ramified primes are of class 1. Hence,

$$h \circ \partial[M,B] = \Big( \prod_{P \text{ inert}} (-1)^{\partial(P)[M,B]} \prod_{P \text{ ramified}} (d,\sigma)_P,\ 1,\ 1,\ \ldots\ 1\Big)$$

However, by Hilbert reciprocity,

$$\prod_{\text{all } P} (d,\sigma)_P = +1. \quad \text{Thus,}$$

$$\prod_{P \text{ inert}} (-1)^{\partial(P)[M,B]} \prod_{P \text{ ramified}} (d,\sigma)_P = \prod_{P \text{ inert}} (-1)^{v_P(I)}$$

as claimed. □

Let $\tilde{h}: H_u(E/I) \to G/W_2$ by the composition of $h: H_u(E/I) \to G$ with projection $G \to G/W_2$.

<u>Corollary 4.4</u> <u>Again with the hypotheses of Lemma</u> 5.1, <u>there is an exact sequence</u>

$$0 \to \text{image } \partial \to H_u(E/I) \xrightarrow{\tilde{h}} G/W_2 \to 0.$$

<u>Proof</u>: By Lemma 4.3, image $\partial$ is contained in the kernel of $\tilde{h}$. It follows by realization of Hilbert symbols and Theorem 3.1 that image $\partial$ = kernel $\tilde{h}$.

$h$ and consequently $\tilde{h}$ maps onto $G$ by realization of Hilbert symbols. □

Corollary 4.4 in fact is the proof of part (2) of the following theorem.

Theorem 4.5 The cokernel of $\partial: H_u(E) \to H_u(E/I)$ is given as follows.

(1) If there are no ramified primes, finite or infinite, the cokernel of $\partial$ depends on the number of primes at which $v_p(x_1) + v_p(I)$ is odd. If this number is even, the cokernel of $\partial$ is $C_2$. If this number is odd, $\partial$ is onto.

(2) If there are no signatures, dyadic ramified primes, and there are ramified primes, all ramified primes being class 1, then there is an isomorphism induced by $\tilde{h}$:

$$H(E(I)/\text{im}\,\partial = \text{cokernel} \quad \partial \simeq G/W_2$$

Thus the cokernel has order $2^{2t + r}$. There is an element of order 4 in cokernel $\partial$ if and only if $t \neq 0$.

(3) If there are signatures, dyadic ramified primes, or ramified primes of class 0 the cokernel of $\partial$ is a product of $C_2$'s. Its order is determined as follows. Let $a$ be the number of ramified primes of class 1. Let $b$ be the number of dyadic ramified primes of class 0. Then the cokernel has order $2^{\max(a - 1,0) + b}$.

Proof: (1) If there are no ramified primes, we must consider two cases.

(a) $v_p(x_1)$ is odd at an even number of primes.

(b) $v_p(x_1)$ is odd at an odd number of primes.

In case (a), we can read $\partial: H_u(E) \to H_u(E/I) \simeq \bigoplus_P H(D/P)$ by Theorem 3.1. The formulas for $\partial(P)$ are:

$$(a,\sigma)_P = (-1)^{\partial(P)\langle ax_1\rangle + v_P(I)} \qquad \text{on rank 1 forms } \langle ax_1\rangle.$$

$$(d,\sigma)_P = (-1)^{\partial(P)(B)} \qquad d = \text{discriminant of } B, \text{ on even rank forms.}$$

Suppose $v_P(I)$ is odd at an even number of primes. Then clearly $v_P(x_1) + v_P(I)$ is odd at an even number of primes, since we are in case (a).

By realization of Hilbert symbols we may pick $a \in F$ with $(a,\sigma)_P$ arbitrarily specified subject only to reciprocity. Since $v_P(I)$ is odd at an even number of primes, the formulas given determine that $\partial(P)$ is non-trivial at an even number of primes. Thus the cokernel has order 2 by realization; and $\partial$ is subject only to the stated restriction.

Continuing, if $v_P(I)$ is odd at an odd number of primes, $v_P(x_1) + v_P(I)$ is odd at an odd number of primes also. Thus by the formula $(a,\sigma)_P = (-1)^{\partial(P)\langle ax_1\rangle + v_P(I)}$, we may make $\partial(P)$ non-trivial at any one specified prime, and $\partial$ is onto.

(b) $v_P(x_1)$ is odd at an odd number of primes. In this case, by Theorem 3.1, we have the previously stated formulas, with one exception $P_1$ at which the Hilbert symbol is read "backwards."

Again, suppose $v_P(I)$ is odd at an odd number of primes. This means $v_P(x_1) + v_P(I)$ is odd at an even number of primes. However, reciprocity now reads

$$\prod_{P} (a,\sigma)_{P} = \prod_{P \neq P_1} (-1)^{\partial(P)\langle ax_1\rangle + v_P(I)} (-1)^{\partial(P_1)\langle ax_1\rangle + v_{P_1}(I) + 1}$$

from which it becomes clear that $\partial(P)$ must be non-trivial at an even number of primes as before. The last case is also similar, and the boundary is onto when $v_P(x_1) + v_P(I)$ is odd at an odd number of primes.

<u>Comment</u>: In this instance we are penalized for our choice of $x_1$. Had we chosen $x_1$ with $v_P(x_1) \equiv v_P(I) \pmod 2$ almost everywhere, (meaning with at most one exception) the local boundary $\partial(P)$ would have given the formula $(a,\sigma)_P = (-1)^{\partial(P)\langle ax_1\rangle}$ almost everywhere, depending on the number of primes at which $v_P(x_1) + v_P(I)$ is odd. Part (1) of this theorem would have required no special analysis. Indeed, since the cokernel of $\partial$ does not depend on $x_1$, this provides an alternate proof.

However, we find it more natural to give $x_1$ with even valuation at inerts. Thus, forms in $H_{+1}(E)$ do not require a special "$x_1$" in their diagonalization.

(2) As remarked earlier, Part (2) follows from Corollary 4.4.

(3) If there are signatures, dyadic ramified primes or tamely ramified primes of class 0, Hilbert symbols determining boundary may be arbitrarily specified, with corrections made at the infinite ramified, dyadic or class 0 primes. Thus, by realization there is a form in $H(E)$ with non-trivial discriminant at prescribed inert and ramified primes of class 1. Applying $\partial$, we obtain a form in $H(E/I)$ with prescribed rank at inert primes, and prescribed discriminant at class 1 ramified primes. However, as required by Theorem 3.1, rank is

preserved at class 1 ramified primes, with $\partial(P) = 0$ at dyadic ramified primes of class 0.

We may thus write

$$H(E/I)/\mathrm{im}\,\partial \cong (F_2 \times \ldots \times F_2)^a \times (F_2 \ldots \times F_2)^b / W \quad \text{via } \gamma$$

with $\gamma$ given by:

$$\gamma : \oplus [M_i, B_i] \to (r_1, \ldots r_a, s_1, \ldots s_b) \quad \text{where}$$

$r_i$ = rank $[M_i, B_i]$ at class 1 ramified primes, dyadic or not.
$s_i$ = rank $[M_i, B_i]$ at class 0 dyadic ramified primes.
$W = C_2$ subgroup of the product

$$\prod_{i=1}^{a+b} F_2 \quad \text{generated by} \quad \begin{matrix}(1, \ldots 1, 0, 0, \ldots 0) \\ (0, \ldots 0, 0, \ldots 0)\end{matrix}$$

This completes the proof. □

<u>Comment</u>: $\partial(P) = 0$ at tamely ramified class 0 primes and also $H_u(D/P) = 0$. Thus, there is no contribution to the cokernel.

By applying Theorem 3.1 we are also able to calculate $H_u(I)$.

We let $J \subset H(E)$ be the subgroup of even rank forms. On $J \cap H_u(I)$ we may define a local discriminant homomorphism at each dyadic ramified prime:

$$d_i\colon J \quad H_u(I) \;\rightarrow\; H(E) \;\xrightarrow{(d,\sigma)_{P_i}}\; Z^*$$

where $d$ is the discriminant of the form in $H(E)$, and $(d,\sigma)_{P_i}$ is the Hilbert symbol at $P_i$. Since we have restricted the domain of $d_i$ to the even rank forms, the discriminant is multiplicative, as is the Hilbert symbol, so that $d_i$ is indeed a homomorphism.

Let $t$ be the number of dyadic ramified primes. We then define the total discriminant homomorphism $\tilde{d}$ to be the product of the $d_i$.

$$\tilde{d} = \Pi(d_i) : J \cap H_u(I) \rightarrow (Z^*)^t.$$

There are also the infinite ramified primes to take care of. By Chapter II 5, we may define a signature $sgn_i\colon H_u(E) \rightarrow Z$ at each infinite ramified prime $P^i_\infty$. Combining, we obtain a total signature homomorphism $sgn\colon J \cap H(I) \rightarrow (2Z)^r$, where $r$ is the number of infinite ramified primes.

Recall how the Hilbert symbol was read at an infinite ramified prime, Chapter II 2.5. Let $d$ be the discriminant, $P_\infty$ an infinite ramified prime. Then $(d,\sigma)_\infty = \pm 1$ depending on the sign of $d$ in $P_\infty$. $d = (-1)^{w(w-1)/2}\det$ where $w = \text{rank } B$, $\det = \text{determinant } B$. Hence $d$ has signature

$$(-1)^{\frac{w(w-1)}{2}}\,(-1)^{\frac{w-\text{sgn}B}{2}} = (-1)^{\frac{w^2-2w-\text{sgn}B}{2}}.$$

For $[M,B]$ in $J \cap H_u(I)$, $w \equiv 0 \mod 2$, so that

$$(d,\sigma)_\infty = (-1)^{-\text{sgn}B/2} = (-1)^{\text{sgn}B/2}.$$

We thus may state:

Theorem 4.6 $H_u(I)$ is determined as follows. For the even rank forms in $H_u(I)$, we have an exact sequence:

$$0 \to J \cap H_u(I) \xrightarrow{\operatorname{sgn} \oplus \tilde{d}} (2Z)^r \oplus (Z^*)^t \xrightarrow{H} Z^* \to 0$$

where $H = \prod_{i=1}^{r} (-1)^{\operatorname{sgn}_i/2} \prod_{i=1}^{t} d_i$ is the Hilbert reciprocity map.

In order for a rank 1 form to exist in $H_u(I)$ there must be no class 1 ramified primes. If there are any signatures or class 0 ramified primes, a rank 1 form exists. If there are no ramified primes, a rank 1 form exists if and only if $v_P(I) + v_P(x_1) \equiv 1 \ (2)$ at an even number of primes.

Further, if a rank 1 form exists, there is an exact sequence:

$$0 \to J \cap H_u(I) \to H(I) \xrightarrow{\text{rank}} F_2 \to 0$$

This sequence splits if and only if $-1$ is a norm.

Proof: $J \cap H(I)$ embeds into $J \cap H(E)$, which by Landherr's Theorem II 5.4 is determined by the discriminant and multisignatures. However, $H_u(I) = \bigcap_{P = \overline{P}} \ker \partial(P)$, by Corollary 2.8. Now $\partial = \oplus \, \partial(P)$ is read by the discriminant at all inert primes of even rank, and at all class 1 tamely ramified primes. Hence $(d,\sigma)_P$ must be trivial at all these primes in order for a form to be in the kernel of $\partial = \oplus \, \partial(P)$.

Thus, by Landherr we obtain

$$0 \to J \cap H_u(I) \xrightarrow{\text{sgn} \oplus \tilde{d}} (2Z)^r \oplus (Z^*)^t \xrightarrow{H} Z^* \to 0$$

is exact.

To complete the discussion, we must examine whether a rank 1 form can exist in $H_u(I)$. $\langle ax_1\rangle$ is such a form if and only if $\partial(P)\langle ax_1\rangle = 0$. By Theorem 3.1, this is possible only if there are no class 1 ramified primes. Assuming this necessary condition, if there is a signature or a class 0 ramified prime, we apply Realization of Hilbert symbols. Thus, in this case, there exists $a \in F/N_{E/F}$ with $\partial(P)\langle ax_1\rangle = 0$ at all inert primes. Further, all ramified primes are of class 2, so that $\partial\langle ax_1\rangle = 0$, and we obtain the desired rank 1 form.

We apply Theorem 3.1 if there are no ramified primes, finite or infinite. In order to have $\partial(P)\langle ax_1\rangle = 0$ we must satisfy:

$$(a,\sigma)_P = (-1)^{\partial(P)\langle ax_1\rangle + v_P(I)} = (-1)^{v_P(I)}$$

when $v_P(x_1)$ is odd at an even number of primes. This is possible if and only if $v_P(I)$ is odd at an even number of primes, so that $v_P(I) + v_P(x_1)$ is odd at an even number of primes.

Similarly, if $v_P(x_1)$ is odd at an odd number of primes, $\langle ax_1\rangle$ must satisfy:

$$(a,\sigma)_P = (-1)^{\partial(P)\langle ax_1\rangle + v_P(I)} = (-1)^{v_P(I)} \qquad P \neq P_1$$

$$(a,\sigma)_{P_1} = (-1)^{v_P(I)+1}$$

in order to have $\partial\langle ax_1\rangle = 0$. Again Hilbert reciprocity implies $v_P(I)$ is odd at an even number of primes, so that $v_P(I) + v_P(x_1) \equiv 1 \pmod 2$ at an even number of primes.

Finally, we must discuss the extension

$$0 \to J \cap H_u(I) \to H(I) \xrightarrow{rk} F_2 \to 0$$

when a rank 1 form exists. By Landherr, and Corollary II 4.14, this sequence splits if and only if $-1$ is a norm in F/NE. □

In this Theorem we have computed $H_u(I)$. We should note that when there are no rank 1 forms it is still possible to describe exactly what an additive set of generators for $H_u(I)$ looks like, see [W].

<u>Corollary 4.7</u> <u>If there are no ramified primes, the boundary sequence depends on the number of primes at which</u> $v_P(x_1) + v_P(I)$ <u>is odd</u>.

(a) <u>If this number is even, we have:</u>

$$0 \to H_u(I) \to H_u(E) \to H_u(E/I) \to F_2 \to 0$$

(b) <u>If this number is odd</u>, $H_u(I) = 0$, <u>and</u>

$$0 \to H_u(E) \to H_u(E/I) \to 0. \qquad \square$$

Thus we have shown how to compute $H_u(I)$, and the boundary homomorphism $\partial: H_u(E) \to H_u(E/I)$. To apply this to $I = \Delta^{-1}(D/Z)$ we clearly only need to know $v_P(\Delta^{-1}(D/Z))$. We can read this for dyadic ramified primes in terms of type as we shall see in 4.12.

Let $P$ be a - invariant prime ideal in $O(E)$. Then the - involution makes the local units in $O_E(P)$, denoted $O_E(P)^*$, into a $C_2$-module. We examine the resulting cohomology group $H^1(C_2;O_E(P)^*)$.

Lemma 4.8 If $P = \bar{P}$ is over inert, then $H^1(C_2;O_E(P)^*) = 1$. If $P = \bar{P}$ is over ramified, $H^1(C_2;O_E(P)^*) = C_2$.

Proof: Recall that $H^1(C_2;O_E(P)^*) = \{x \in O_E(P)^*: x\bar{x} = 1\}$ modulo $\{v/\bar{v}: v \in O_E(P)^*\}$. If $x$ is a local unit of norm 1, then by Hilbert Theorem 90, there exists $z \in E^*$ with $z\bar{z}^{-1} = x$. We now write $z$ as $z = \pi^{v_P(z)}v$ where $\pi$ is a local uniformizer for $P$ and $v \in O_E(P)^*$.

If $P$ is over inert, we may choose $\pi \in F^*$, so that $\pi = \bar{\pi}$. Thus $z\bar{z}^{-1} = v\bar{v}^{-1} = x$ and $H^1$ is trivial in this case.

If $P$ is over ramified, $(\pi\bar{\pi}^{-1})$ is a local unit, and $z\bar{z}^{-1} = (\pi\bar{\pi}^{-1})^{v_P(z)}v\bar{v}^{-1} = x$. Thus $H^1$ is generated by the class of $\pi\bar{\pi}^{-1}$ which we denote $cl(\pi\bar{\pi}^{-1})$. Of course $(\pi\bar{\pi}^{-1})^2$ is trivial in $H^1$ since $(\pi\bar{\pi}^{-1})^2 = (\pi\bar{\pi}^{-1})(\bar{\pi}\pi^{-1})^{-1}$ a quotient of local units. Thus, to complete the proof we need only show that $cl(\pi\bar{\pi}^{-1})$ is non-trivial in $H^1$.

Suppose to the contrary that there is a local unit $v$ with $v\bar{v}^{-1} = \pi\bar{\pi}^{-1}$. Then $\pi v^{-1} = \overline{\pi v}^{-1}$ so that $\pi v^{-1}$ is a local uniformizer of $O_E(P)$ which lies in $F$. This is impossible as we are in the ramified case. Hence, $cl(\pi\bar{\pi}^{-1})$ is non-trivial as claimed. □

Definition 4.9 $P$ is of type 1 provided $cl(-1) \neq 1 \in H^1(C_2;O_E(P)^*)$. $P$ is of type 2 provided $cl(-1) = 1 \in H^1(C_2;O_E(P)^*)$.

We can thus classify type by:

Proposition 4.10 $P$ is of type 1 if and only if there exists a local uniformizer $\pi \in O_E(P)$ with $\bar{\pi} = -\pi$. $P$ is of type 2 if and only if there is a skew unit, $u = -\bar{u} \in O_E(P)^*$.

Proof: $cl(-1) \neq 1$ if and only if for any local uniformizer $\pi$, $cl(-1) = cl(\pi\bar{\pi}^{-1})$. Hence, for some $v \in O_E(P)^*$, $-v\bar{v}^{-1} = \pi\bar{\pi}^{-1}$. Replacing $\pi$ by $\pi\bar{v}$, we obtain a skew uniformizer. $cl(-1) = 1$ if and only if there is a local unit $v$ with $v\bar{v}^{-1} = -1$. □

Proposition 4.11 If $P$ is non-dyadic ramified, meaning the characteristic of the residue field $O(F)/P \neq 2$, then $P$ is of type 1.

Proof: Since $P$ is ramified, $e = 2$ and $f = 1$. Thus $O(E)/P = O(F)/P$, and the induced involution on $O(E)/P$ is trivial.

Suppose $P$ is of type 2. Then by 4.10, there is a skew unit $v$, with $v = -\bar{v}$. However, viewing $v$ in the residue field, $v = \bar{v}$ Thus $1 = -1$. This is a contradiction unless the characteristic is 2. □

Finally, we wish to examine type in terms of the local different. Let $P$ ramify, say $PO(E) = P^2$. By [A 83], we can find $\alpha \in \tilde{O}_E(P)$ where $\tilde{O}_E(P)$ denotes the localized completion of $O(E)$ at $P$, such that $1, \alpha$ forms a basis of $O_E(P)$, as an $O_F(P)$ - module. $\alpha$ satisfies $t^2 - (\alpha + \bar{\alpha})t + \alpha\bar{\alpha} = 0$. By [A 92], $2\alpha - (\alpha + \bar{\alpha}) = \alpha - \bar{\alpha}$ generates the different of the extension.

Factor $\alpha - \bar{\alpha} = \pi^{d_P} v$, where $d_P$ is called the local differential expotent, $\pi$ is a local uniformizer for $P$, $v \in \tilde{O}_E(P)^*$.

We have

$$\overline{(\alpha - \bar{\alpha})} = -(\alpha - \bar{\alpha}) = \bar{\pi}^{d_P}\bar{v} = -\pi^{d_P} v.$$

Thus, $-v\bar{v}^{-1} = (\bar{\pi}\pi^{-1})^{d_P}$ and $cl(-1) = cl(\pi\bar{\pi}^{-1})^{d_P}$. Thus, by definition,

Proposition 4.12 Let P be dyadic ramified. Then P is of type 1 if and only if the local differential exponent $d_P$ is odd. □

## Chapter VII NON-MAXIMAL ORDERS

In this chapter we discuss the ingredients necessary to compute $W(k,Z;S)$ for $S$ a non-maximal order.

First there is the problem of the map $\tilde{w}: O(E)/P \to E/\Delta^{-1}(D/Z)$ from Chapter VI 3. $\tilde{w}$ is determined by a localizer $\rho$ satisfying $v_P(\rho) = v_P(\Delta^{-1}(E/Q)) - 1$. However, we must choose a canonical localizer $\rho(P)$ in order that the diagram given after Theorem VI 2.5 will commute, and be applicable to our computation.

A series of theorems of Conner develop the fundamental properties of these canonical localizers. In Section 2, normal extensions are discussed.

The basic question is for a non-dyadic ramified prime $P$ whether $\rho(P)$ is a local norm. The best result known is the result of Conner, Theorem 2.9, which shows that in some cases it is equivalent to asking whether $1/p$ is a local norm.

In Section 3 we compute the trace map for finite fields.

In Section 4 we discuss the conductor and the set $T(M) = \{$maximal ideals $P$ in $D$: $P \cap S = M$, $P = \overline{P}\}$.

As previously noted, these are the three crucial steps to understanding non-maximal orders.

### 1. Traces and canonical localizers

In the previous chapter we have studied the boundary map

$$\partial(D,P): H_u(E) \xrightarrow{\partial(D)} H_u(E/\Delta^{-1}(D/Z)) \xrightarrow{q(\rho)} H_u(D/P)$$

The manner in which this map is read depends on the choice of localizer $\rho(P)$, with $v_P(\rho) = v_P(\Delta^{-1}) - 1$, which embeds $D/P$ into $E/\Delta^{-1}(P)$. Our object is to study non-maximal orders. To do this requires choosing the isomorphism $H_u(E/\Delta^{-1}(D/Z)) \cong \bigoplus_{P=\overline{P}} H_u(D/P)$ in such a way that the following diagram commutes:

$$\begin{array}{ccc} H_u(E/\Delta^{-1}(D/Z)) & \dashrightarrow & \bigoplus H_u(D/P) \\ \downarrow t & & \downarrow tr \\ W(k,Q/Z;D) & \xrightarrow{g} & \bigoplus_{P=\overline{P}} W(k,F_p;D/P) \end{array}$$

Thus we are faced with the problem of relating the number field trace $t$ to the residue field trace $tr$. Again we recall the setting in which we work.

We let $E$ be an algebraic number field, with $F \subset E$ a subfield. $O(E) \subset E$ and $O(F) \subset F$ are the respective Dedekind rings of algebraic integers. If $P$ is a prime ideal in $O(E)$, then $P \cap O(F) = \mathrm{P}$ is the corresponding prime ideal in $O(F)$ over which $P$ lies.

Consider the quotient homomorphism $\gamma_P: O(E) \to O(E)/P$ onto the finite residue field. Restricting $\gamma_P$ to $O(F)$ induces $\gamma_{\mathrm{P}}: O(F) \to O(F)/\mathrm{P} \to O(E)/P$.

Our object is to compare the two trace homomorphisms:

$$t_{E/F}: E \to F \quad \text{and} \quad tr_{E_P/F_{\mathrm{P}}} : O(E)/P \to O(F)/\mathrm{P}$$

where $E_P$ is an abbreviation for $O(E)/P$ the residue field.

We consider the diagram:

$$\begin{array}{ccc} O(E) & \overset{\tau}{- - - \to} & O(F) \\ \downarrow \gamma_{\mathcal{P}} & & \downarrow \gamma_P \\ O(E)/\mathcal{P} & \overset{tr}{\to} & O(F)/P \end{array}$$

The composition $tr \circ \gamma_{\mathcal{P}}$ is an $O(F)$ - module epimorphism, as is $\gamma_P$. Since $O(E)$ is projective as an $O(F)$ - module, there is an $O(F)$ - module homomorphism $\tau: O(E) \to O(F)$ which makes the diagram above commute.

Now consider the relative inverse different $\Delta^{-1}(E/F) \simeq \mathrm{Hom}_{O(F)}(O(E), O(F))$ [A 86]. Since $\tau \in \mathrm{Hom}_{O(F)}(O(E), O(F))$, using this isomorphism we see that there is an element

$$\mu_{E/F}(\mathcal{P}) \quad \varepsilon \quad \Delta^{-1}(E/F)$$

such that $t_{E/F}(\mu_{E/F}(\mathcal{P})\lambda) = \tau(\lambda)$. Thus,

$$(*) \quad \gamma_P(t_{E/F}(\mu_{E/F}(\mathcal{P})\lambda)) = tr(\gamma_{\mathcal{P}}(\lambda)) \quad \text{for all} \quad \lambda \in O(E).$$

Now if $\mu^1_{E/F}(\mathcal{P})$ and $\mu^2_{E/F}(\mathcal{P})$ are two elements satisfying (*), then clearly $t_{E/F}((\mu^1 - \mu^2)\lambda) \in P$ for all $\lambda \in O(E)$. Thus, using the isomorphism

$$P\Delta^{-1}(E/F) \simeq \mathrm{Hom}_{O(F)}(O(E), P)$$

it follows that $\mu^1 - \mu^2 \in P\Delta^{-1}(E/F)$.

So although $\mu_{E/F}(P)$ is not unique in $\Delta^{-1}(E/F)$, it is unique in $\Delta^{-1}(E/F)/P\Delta^{-1}(E/F)$.

The following Lemmas of Conner develop the basic properties of these $\mu_{E/F}(P)$.

Lemma 1.1 For any $\mu_{E/F}(P)$, $\mu_{E/F}(P)PP^{-1} \subset \Delta^{-1}(E/F)$.

Proof: Let $\alpha \in P$ and $\beta \in P^{-1}$. We abbreviate $\mu_{E/F}(P)$ by $\mu$. Then for any $\lambda \in O(E)$

$$t_{E/F}(\lambda\ \mu\ \alpha\ \beta) = \beta\ t_{E/F}(\lambda\ \alpha\ \mu)$$

since $\beta \in P^{-1} \subset F$. But $t_{E/F}(\lambda\ \alpha\ \mu) \in P$ since $\gamma_P(\lambda\ \alpha) = 0$ and $\gamma_P(t_{E/F}(\lambda\ \alpha\ \mu)) = tr(\gamma_P(\lambda\ \alpha))$. Hence $\beta\, t_{E/F}(\lambda\ \alpha\ \mu) = t_{E/F}(\lambda\ \mu\ \alpha\ \beta) \in O(F)$ for all $\lambda \in O(E)$. Thus by definition of the inverse different, $\mu\ \alpha\ \beta \in \Delta^{-1}(E/F)$. □

Lemma 1.2 Let $(\mu_{E/F}(P))$ be the principal $O(E)$ - ideal generated by $\mu$. Then $\mu P^{-1}$ is not contained in $\Delta^{-1}(E/F)$.

Proof: Suppose the $\mu P^{-1} \subset \Delta^{-1}(E/F)$. Then for all $\beta \in P^{-1}$ and $\lambda \in O(E)$

$$t_{E/F}(\lambda\ \mu\ \beta) = \beta\ t_{E/F}(\lambda\ \mu) \in O(F)$$

So $t_{E/F}(\lambda\ \mu) \in (P^{-1})^{-1} = P$. However, we also have

$$\gamma_P t_{E/F}(\lambda\mu) = tr_{E_P/F_P}(\gamma_P(\lambda)), \text{ so } tr_{E_P/F_P} : E_P \to F_P$$

is trivial. Contradiction. □

When $F = Q$ we may use these $\mu$ as follows. For each prime ideal $P \subset O(E)$ we obtain a canonically defined element

$$\rho(P) = p^{-1}\mu_{E/Q}(P) \in E/\Delta^{-1}(E/Q)$$

where $p \in Z$ is the prime over which $P$ lies. Now define a homomorphism

$$O(E) \to E/\Delta^{-1}(E/Q) \quad \text{by} \quad \lambda \to (\lambda\rho(P)).$$

By Lemma 1.1 this map induces a map $\tilde{w}$ of the residue fields $\tilde{w}: O(E)/P \to E/\Delta^{-1}(E/Q)$ since $\mu_{E/Q}(P)P\,p^{-1} \subset \Delta^{-1}(E/Q)$. By Lemma 1.2, the induced map is an embedding since $\mu_{E/Q}(P)p^{-1} \not\subset \Delta^{-1}(E/Q)$.

We also have the map $Z \to Q/Z$ defined by $n \to (np^{-1})$. This also induces an embedding $w$: $F_p = Z/pZ \to Q/Z$. Combining $w$ and $\tilde{w}$ we have

<u>Theorem 1.3</u> <u>At every prime</u> $P \subset O(E)$ <u>there is a commutative diagram</u>

$$\begin{array}{ccccc} 0 \to & E_P = O(E)/P & \xrightarrow{\tilde{w}} & E/\Delta^{-1}(E/Q) \\ & \downarrow tr & & \downarrow t \\ 0 \to & F_p & \xrightarrow{w} & Q/Z \end{array}$$

$\tilde{w}$ and $w$ have been described.

tr = finite field trace

t = map induced from number field trace $t_{E/Q}$.

<u>Proof</u>: This follows immediately by the definition of $\mu_{E/Q}(P)$, since

$$\tilde{w}(\lambda) = \lambda\rho(P) = \lambda\rho^{-1}\mu_{E/Q}(P), \text{ so}$$

$$t(\tilde{w}(\lambda)) = \frac{1}{p}\, t\,(\lambda\mu_{E/Q}(P))$$

$$= \frac{1}{p}\, tr(\lambda) = w(tr(\lambda)). \quad \square$$

Thus we are able to explicitly describe the map $\tilde{w}$ given in Chapter VI 3. As we saw there, the canonical localizer $\rho(P)$ which defines $\tilde{w}$ is the correct choice to make if one wishes to compute $W(k,Z;S)$, where $S$ is a non-maximal order. The reader is referred to the discussion preceding Proposition VI 2.7.

The choice we make for localizer $\rho(P)$ will determine the isomorphism

$$H(E/\Delta^{-1}(E/Q)) \simeq \bigoplus_{P=\bar{P}} H(E/\Delta^{-1}(E/Q)(P))$$

$$\simeq \bigoplus_{P=\bar{P}} H(E_P)$$

since the embedding $E_P \to E/\Delta^{-1}(E/Q)(P)$ defines the isomorphism $H(E_P) \to H(E/\Delta^{-1}(E/Q)(P))$.

There are two cases.

(1) If $P$ is inert, the isomorphism is independent of localizer, and rank mod 2 is the only invariant.

(2) If $P$ is ramified, $W(E_P) \approx H(E_P) \approx H(E/\Delta^{-1}(E/Q)(P))$ since the induced involution is trivial. For $P$ non-dyadic, the identification does depend on the choice of localizer.

The localizer we choose therefore will determine the way we read the boundary homomorphism $\partial(P): H(E) \to H(E_P)$.

We wish to compare the canonical localizer with the localizer given in Theorem VI 3.1. Now different localizers $\rho, \rho_1$ determine different embeddings $E_P \to E/\Delta^{-1}(E/Q)(P)$. The relation between different embeddings is that one is the composition of the other with the map of $E_P$ to itself given by multiplication by $\rho\rho_1^{-1}$. Since we are only interested in this map on $H(E_P)$, only the norm class of $\rho\rho_1^{-1}$ matters.

For the localizer in Theorem VI 3.1, this norm class is specified. Thus, in order to understand the canonical localizer above, the question we need to answer is whether $\rho(P)$ is a local norm. Once we have this information, we can read the boundary homomorphism $\rho(P)$ with Hilbert symbols, where the identifications come from our canonical localizers.

Since $\rho(P) = \rho^{-1}\mu_{E/Q}(P)$, we must study those elements $\mu_{E/Q}$. The following theorems of Conner are fundamental.

Theorem 1.4 For any $\mu_{E/F}(P)$ we have:

(1) $v_P(\mu_{E/F}(P)) = v_P(P) - 1 + v_P(\Delta^{-1}(E/F))$

(2) If $\tilde{P} \neq P$ then $v_{\tilde{P}}(\mu_{E/F}(P)) \geq v_{\tilde{P}} P + v_{\tilde{P}}(\Delta^{-1}(E/F))$.

Proof: By 1.1, $\mu PP^{-1} \subset \Delta^{-1}(E/F)$. Thus

$$v_{\tilde{P}}(\mu) + v_{\tilde{P}}(P) - v_{\tilde{P}}(P) \geq v_{\tilde{P}}\Delta^{-1}(E/F).$$

If $\tilde{P} \neq P$ then $v_{\tilde{P}}(P) = 0$, and (2) follows.

Again using 1.1, we have $v_P(\mu) + 1 - v_P(P) \geq v_P\, \Delta^{-1}(E/F)$. However, if $v_P(\mu) - v_P(P) \geq v_P\, \Delta^{-1}(E/F)$, then $\mu P^{-1} \subset \Delta^{-1}(E/F)$ which contradicts 1.2. This shows equality and proves (1).

We next recall some facts about the different $\Delta(E/F)$, and its inverse $\Delta^{-1}(E/F)$. Let $\tilde{e} > 0$ be defined by $v_{\tilde{P}}(P) = \tilde{e}$.

Definition 1.5 If $v_{\tilde{P}}(P) = \tilde{e} > 0$ is divisibly by the characteristic of the residue field $ch(F_P)$ then we say $\tilde{P}$ is wildly ramified over $P$.

We recall that if $\tilde{P}$ is wildly ramified $v_{\tilde{P}}(\Delta(E/F)) \geq \tilde{e}$. [S 96].

Definition 1.6 If $v_{\tilde{P}}(P) = \tilde{e} > 0$ is not divisible by $ch(F_P)$ then we say $\tilde{P}$ is tamely ramified over $P$.

In this case $v_{\tilde{P}}(\Delta(E/F)) = \tilde{e} - 1$. Thus $v_{\tilde{P}}\Delta(E/F) = \tilde{e} - 1$ if and only if $\tilde{P}$ is tamely ramified.

Definition 1.7 A prime $P \subset O(F)$ is tame in $E/F$ if and only if no prime $\tilde{P} \subset O(E)$ which divides $PO(E)$ is wildly ramified.

Thus $P$ is tame if and only if for every prime $\tilde{P}$ with $v_{\tilde{P}}P = \tilde{e} > 0$, $\tilde{e}$ is relatively prime to $ch\ F_P$.

Theorem 1.8 Let $\mu_{E/F}(P)$ satisfy

$$\gamma_P t_{E/F}(\mu_{E/F}(P) \cdot \lambda) = tr_{E_P/F_P}(\gamma_P(\lambda))$$

(1) If $v_P(P) = e > 0$ is relatively prime to $ch(F_P)$, then $\mu_{E/F}(P) \in O_E(P)^*$ is a local unit.

(2) If $v_P(P) = e > 0$ is divisible by $ch(F_P)$, then $\mu_{E/F}(P)$ is not a local integer at $P$.

(3) If $\tilde{P} \neq P$ but $v_{\tilde{P}}(P) = \tilde{e} > 0$ is relatively prime to $ch(F_P)$ then $\mu_{E/F}(P)$ belongs to the maximal ideal $m(\tilde{P}) \subset O_E(\tilde{P})$.

(4) If $\tilde{P}$ is prime to both $PO(E)$ and $\Delta(E/F)$ then $\mu_{E/F}(P)$ is a local integer in $O_E(\tilde{P})$.

Proof: We use the previous remarks about the inverse different together with Theorem 1.4.

(1) By 1.4, $v_P(\mu) = (e - 1) + -(e - 1) = 0$, so $\mu$ is a local unit.

(2) In this case $v_P(\Delta^{-1}(E/F)) \leq -e$. Thus by 1.4, $v_P(\mu) \leq (e - 1) - e = -1$, so $\mu$ is not a local integer.

(3) By 1.4, $v_{\tilde{P}}(\mu) \geq \tilde{e} - (\tilde{e} - 1) = 1$, so $\mu \in m(\tilde{P})$.

(4) By 1.4, $v_{\tilde{P}}(\mu) \geq 0$ so $\mu \in O_E(\tilde{P})$ is a local integer. □

## 2. Normal extensions

We continue by letting $E/F$ be a normal extension, and we let $G$ be the Galois group. Again, this section is due to Conner.

<u>Theorem 2.1</u> <u>If</u> $g \in G$ <u>and</u> $P \subset O(E)$ <u>is a prime ideal then</u> $g\mu_{E/F}(P) = \mu_{E/F}(gP)$.

<u>Proof</u>: Let $g_*$ be the induced $F_p$ - linear isomorphism induced by $g$ making the diagram below commute.

$$\begin{array}{ccc} O(E) & \xrightarrow{g} & O(E) \\ \downarrow \gamma_P & & \downarrow \gamma_{gP} \\ E_P = O(E)/P & \xrightarrow{g_*} & O(E)/gP = E_{gP} \end{array}$$

For $\lambda \in O(E)$, $t_{E/F}(\lambda g(\mu)) = t_{E/F}(g^{-1}(\lambda)\mu)$. Recall the defining equation for $\mu_{E/F}(gP)$

$$tr_{E_{gP}/F_p}(\gamma_{gP}(\lambda)) = \gamma_p\, t_{E/F}(\lambda\mu_{E/F}(gP)).$$

We also need the fact that $tr_{E_P/F_p}(\alpha) = tr_{E_P/F_p}(\phi\alpha)$

for all $\phi$ in the Galois group $E_P/F_p$. We now compute:

$$\begin{aligned} \gamma_p t_{E/F}(\lambda\, g(\mu_{E/F}(P))) &= \gamma_p t_{E/F}(g^{-1}(\lambda)\mu_{E/F}(P)) \\ &= tr_{E_P/F_p}(\gamma_P(g^{-1}(\lambda))) \\ &= tr_{E_P/F_p}(g_*\gamma_P g^{-1}(\lambda)) \\ &= tr_{E_P/F_p}(\gamma_{gP}(\lambda)) \end{aligned}$$

Thus $g(\mu_{E/F}(P)) = \mu_{E/F}(gP)$. □

Now let $G_{\mathcal{P}}$ denote the decomposition subgroup at $\mathcal{P}$, ie. all $g \in G$ with $g\mathcal{P} = \mathcal{P}$.

__Theorem 2.2__ __If__ $\mathcal{P} \subset O(E)$ __lies over__ $P$, __and__ $P$ __is tame in__ $E/F$, __then for__ $g \notin G_{\mathcal{P}}$, $g\mu_{E/F}(\mathcal{P}) \in m(\mathcal{P})$.

__Proof:__ Since $g \notin G_{\mathcal{P}}$, $g\mathcal{P} \neq \mathcal{P}$. Thus by 2.1, $g\mu_{E/F}(\mathcal{P}) = \mu_{E/F}(g\mathcal{P})$. We now apply 1.8 to conclude that $g\mu \in m(\mathcal{P})$. □

__Theorem 2.3__ __If__ $\mathcal{P} \subset O(E)$ __lies over__ $P$, __and__ $P$ __is tame in__ $E/F$, __then__ $\mu_{E/F}(\mathcal{P}) \in O_E(\mathcal{P})^*$ __and__ $\gamma_{\mathcal{P}}(\mu_{E/F}(\mathcal{P})) \in F_P$.

__Proof:__ By 1.8 (1), $\mu \in O_E(\mathcal{P})^*$. If $g \in G_{\mathcal{P}}$, then by 2.1, $\mu_{E/F}(\mathcal{P}) - g\mu_{E/F}(\mathcal{P}) \in P\Delta^{-1}(E/F)$. However, $P$ is tame so $v_{\mathcal{P}}(P\Delta^{-1}(E/F)) = 1$, since $v_{\mathcal{P}}(P) = e$ and $v_{\mathcal{P}}(\Delta^{-1}(E/F)) = 1 - e$.

$$\begin{aligned} \text{Hence} \quad \gamma_{\mathcal{P}}(\mu_{E/F}(\mathcal{P})) &= \gamma_{\mathcal{P}}(g\mu_{E/F}(\mathcal{P})) \\ &= g_*\gamma_{\mathcal{P}}(\mu_{E/F}(\mathcal{P})) \end{aligned}$$

for all $g \in G_{\mathcal{P}}$. Since $G_{\mathcal{P}}$ maps onto the Galois group of $E_{\mathcal{P}}/F_P$ by $g \to g_*$ it follows that $\gamma_{\mathcal{P}}(\mu_{E/F}(\mathcal{P}))$ is fixed by all $g_*$ in the Galois group of $E_{\mathcal{P}}/F_P$. Thus $\gamma_{\mathcal{P}}(\mu_{E/F}(\mathcal{P})) \in F_P$. □

__Theorem 2.4__ __If__ $\mathcal{P}$ __lies over a prime__ $P$ $O(F)$ __which is tame in__ $E/F$ __then__ $\gamma_{\mathcal{P}}(\mu_{E/F}(\mathcal{P}))e = 1$ __where__ $e = v_{\mathcal{P}}(P)$.

__Proof:__ By definition of trace,

$$t_{E/F}(\lambda\,\mu) \;=\; \sum_{g\,\varepsilon\,G} g(\lambda\,\mu), \quad \text{where} \quad \lambda\ \varepsilon\ O(E).$$

Now $g(\lambda\mu) = g(\lambda)g(\mu)\ \varepsilon\ O_E(P)$ for all $g\ \varepsilon\ G$ by 2.1 and 1.8. However, for all $g \notin G_P$, $g(\lambda)g(\mu)\ \varepsilon\ m(P)$ by 2.2. Thus

$$\begin{aligned}\gamma_P\, t_{E/F}(\lambda\mu) &= \sum_{g\,\varepsilon\,G_P} \gamma_P(g(\lambda)g(\mu)) \\ &= \sum_{g\,\varepsilon\,G_P} g_*(\gamma_P(\mu))g_*(\gamma_P(\mu))\end{aligned}$$

But by 2.3, $\gamma_P(\mu)\ \varepsilon\ F_P$. Thus, the above equals

$$= \gamma_P(\mu) \sum_{g\,\varepsilon\,G_P} g_*(\gamma_P(\lambda)).$$

Now if we consider $G_P \to \text{Galois}\,(E_P/F_P) \to 1$ this has kernel the inertial subgroup whose order is $e > 0$. Thus by the above

$$\gamma_P t_{E/F}(\lambda\mu) \;=\; e\gamma_P(\mu)\,tr_{E_P/F_P}(\gamma_P(\lambda)).$$

However by definition

$$\gamma_P t_{E/F}(\lambda\mu) \;=\; tr_{E_P/F_P}(\gamma_P(\lambda)).$$

Thus $e\gamma_P(\mu) = 1$ as was to be shown. □

<u>Corollary 2.5</u> <u>Suppose</u> $P \subset O(E)$ <u>lies over a prime tame in</u> $E/F$ <u>and suppose</u> $G_P = G$. <u>Let</u> $d$ <u>satisfy</u> $de \equiv 1\ (\text{mod ch}(F_P))$. <u>Then we may choose</u> $\mu_{E/F}(P) = d$.

Proof: We compute

$$\begin{aligned}\gamma_P(t_{E/F}(\lambda d)) &= d\gamma_P(t_{E/F}(\lambda)) \\ &= d\gamma_P(\sum_{g \in G} g(\lambda)) \\ &= d(\sum_{g \in G} g_*\gamma_P(\lambda)) \\ &= de\, tr_{E_P/F_P}(\gamma_P(\lambda)) \\ &= tr_{E_P/F_P}(\gamma_P(\lambda)) \qquad \square\end{aligned}$$

We need one more theorem, which shows that the $\mu$ behave multiplicatively in towers. This makes no assumption of normality.

Theorem 2.6 (Tower Theorem) Let $N \supset E \supset F$, and $\hat{P} \subset O(N)$ a prime ideal with $\hat{P} \cap O(E) = P$, $P \cap O(F) = P$. Then $\mu_{N/F}(\hat{P}) = \mu_{N/E}(\hat{P}) \cdot \mu_{E/F}(P)$.

Proof: We simplify our notation by letting $\mu_1 = \mu_{N/E}(\hat{P})$ and $\mu_2 = \mu_{E/F}(P)$. For $\lambda \in O(N)$ we now compute:

$$\begin{aligned}\gamma_P t_{N/F}(\gamma\mu_1\mu_2) &= \gamma_P t_{E/F}(\mu_2\, t_{N/E}(\lambda\mu_1)) \\ &= tr_{E_P/F_P}(\gamma_P t_{N/E}(\lambda\mu_1)) \\ &= tr_{E_P/F_P}(tr_{N_{\hat{P}}/E_P}(\gamma_{\hat{P}}(\lambda)) \\ &= tr_{N_{\hat{P}}/F_P}(\gamma_{\hat{P}}(\lambda))\end{aligned}$$

Thus, $\mu_1\mu_2 = \mu_{N/F}(\hat{P})$ as desired. □

We recall now that our object has been to determine whether $\rho(P) = p^{-1}\mu_{E/Q}$ is a local norm. In particular, this is important precisely when $P = \overline{P}$ lies over a non-dyadic ramified prime. Here we are letting E have involution -, and the fixed field of - is F.

<u>Corollary 2.7</u> *If $P = \overline{P}$ lies over a non-dyadic ramified prime p then*

$$\mu_{E/Q}(P) = d\,\mu_{F/Q}(P)$$

where $0 < d < ch(F_p) = p$ and $2d \equiv 1 \pmod{p}$.

<u>Proof:</u> By the Tower Theorem 2.6, $\mu_{E/Q}(P) = \mu_{E/F}(P)\mu_{F/Q}(P)$. By Corollary 2.5, $\mu_{E/F}(P) = d$ where $2d \equiv 1 \pmod{p}$. □

<u>Theorem 2.8</u> *If the non-dyadic ramified prime $P \subset O(F)$ lies over a rational prime p which is tame in F/Q then $d\mu_{F/Q}(P)$ is a local norm, ie. the Hilbert symbol $(d\mu_{F/Q}(P),\sigma)_P = 1$ if and only if $2\,v_P(p)$ is a square in the residue field $F_P$.*

<u>Proof:</u> By 2.4 applied to F/Q, $\gamma_P(\mu_{F/Q}(P)) \cdot e = 1$ where $e = v_P(p)$. Also by 2.7, $2d \equiv 1 \pmod{p}$. Thus $2v_P(p)$ is a square in the residue field if and only if its reciprocal $d\gamma_P(\mu_{F/Q}(P))$ is a square. However, by Chapter II 2.4, a local unit is a norm if and only if it is a square in the residue field. □

<u>Theorem 2.9</u> *If the non-dyadic ramified prime* $P \subset O(F)$ *lies over a rational prime* $p$ *which is tame in* $F/Q$ *then* $(\rho(P),\sigma)_P = (1/p,\sigma)_P$ *provided* $\deg F_P/F_p$ *is even.*

<u>Proof</u>: $(\rho(P),\sigma)_P = (1/p,\sigma)_P(d\mu_{F/Q}(P),\sigma)_P$. By Theorem 2.8, $(d\mu_{F/Q}(P),\sigma)_P = 1$ if and only if $2e$ is a square in the residue field $F_P$, where $e = v_P p$. Now consider $2e \in F_p$. If this is already a square then $(d\mu_{F/Q}(P),\sigma)_P = 1$ and we are done. If $2e$ is not a square, then because $\deg F_P/F_p$ is even, $2e$ will be a square in $F_P$, and again $(d\mu_{F/Q}(P),\sigma)_P = 1$. □

Admittedly one still needs criteria to determine whether $1/p$ is a local norm. However, this result of Conner represents a great step foward. What we have attempted to do here is outline the general theory concerning canonical localizers, which are crucial to our understanding of non-maximal orders.

## 3. Computing trace for finite fields

We next compute the trace induced map $tr_*: H(E) \to H(K)$ where $E$ and $K$ are finite fields. As we have observed, this computation is important in computing $W(k,Z;S)$ for non-maximal orders, if we wish to use the commutative diagram after Theorem VI 2.5.

<u>Theorem 3.1</u> *Suppose* $[E : K] < \infty$ *is an extension of finite fields.* $E$ *has an involution - which is possibly trivial. Then the homomorphism* $tr_*$ *is given as follows:*

(1) <u>If</u> - <u>is</u> <u>non-trivial</u> <u>on</u> <u>both</u> E <u>and</u> K,

$tr_*$: H(E) → H(K) <u>is</u> <u>an</u> <u>isomorphism</u>.

(2) <u>If</u> - <u>is</u> <u>trivial</u> <u>on</u> <u>both</u> E <u>and</u> K,

$tr_*$: W(E) → W(K) <u>is</u> <u>an</u> <u>isomorphism</u> <u>if</u> [E : K] <u>is</u> <u>odd</u>.

<u>If</u> [E : K] <u>is</u> <u>even</u>, $tr_*$ <u>is</u> <u>an</u> <u>isomorphism</u> <u>on</u> <u>the</u> <u>fundamental</u> <u>ideal</u>, <u>and</u> <u>has</u> <u>kernel</u> $C_2$.

(3) <u>If</u> - <u>is</u> <u>non-trivial</u> <u>on</u> E, <u>trivial</u> <u>on</u> K, <u>then</u>

$$tr_*: H(E) \to W(K) \ \underline{is} \begin{cases} 1-1 & \underline{if} \ p \neq 2 \\ 0 & \underline{if} \ p = 2. \end{cases}$$

<u>where</u> p <u>is</u> <u>the</u> <u>characteristic</u> <u>of</u> <u>the</u> <u>finite</u> <u>field</u> E.

<u>Proof</u>: We should observe that since E is finite, K is automatically - invariant, and thus has an involution induced on it. Thus the statements make sense.

(1) If [E : K] is even, then K is contained in the fixed field of - by Galois theory, since the fields are finite. Hence in case (1) we may assume [E : K] is odd.

We recall that the Hermitian group of a finite field is determined by rank mod 2 [M,H 117]. So let [M,B] ε H(E). Then $tr_*$[M,B] has rank equal to [rank M] · [E : K]. Hence rank modulo 2 is preserved when [E : K] is odd, so that $tr_*$ is an isomorphism in this case as claimed.

(2) Let [E : K] be odd, with - trivial on E. By [Lm 193], the composition

$$W(K) \xrightarrow{\otimes_K E} W(E) \xrightarrow{tr_*} W(K)$$

is multiplication by $tr_*\langle 1\rangle_E$, which is of odd rank, hence a unit in W(K). Thus the composition is an isomorphism. Since all groups have order 4, $tr_*$ must be an isomorphism too.

Now consider the case $[E : K]$ is even, and $-$ is trivial on $E$. By Galois theory, E/K factors into a maximal odd order extension and successive extensions of degree 2.

$$E/K:\ K \subseteq K_1 \subseteq K_2 \subseteq \ldots \subseteq E$$

$$[K_1 : K] = \text{odd} \qquad [K_i, K_{i-1}] = 2 \quad \text{for} \quad i > 1$$

$$tr_*: W(E) \to W(K) \quad \text{is the composition of the trace maps:}$$

$$W(E) \overset{tr_*}{\to} W(K_i) \to W(K_{i-1}) \to \ldots \to W(K_2) \to W(K_1) \overset{tr_*}{\to} W(K)$$

To finish the proof of (2), we need to examine separately each $tr_*: W(K_i) \to W(K_{i-1})$ and show that each $tr_*$ is an isomorphism on the fundamental ideal, with kernel $C_2$ when $[K_i : K_{i-1}] = 2$. $tr_*: W(K_1) \to W(K)$ is an isomorphism by the first part of the theorem for $[K_1 : K] = \text{odd}$.

Note: We assume now that the characteristic of $E$ is not 2, for in that case, rank mod 2 determines everything, and (2) is clearly true as stated.

We thus consider $tr_*: W(F(\sqrt{w})) \to W(F)$. Since the characteristic of $E$ is not 2, any quadratic extension of $F$ is given by adjoining $\sqrt{w}$, where $w \notin F^{**}$.

In order to describe a basis for $G_{F(\sqrt{w})} \equiv F(\sqrt{w})^*/F(\sqrt{w})^{**}$ as an $F_2$ - vector space, we recall the general theory from [G,F]. Let $G_F = F^*/F^{**}$ have a basis $w, b_1, \ldots, b_n$ as an $F_2$ - vector space.

Then $G_{F(\sqrt{w})}$ has a basis given by $\{b_1,\ldots,b_n\}$ together with $\{x_i + y_i\sqrt{w}\}$ as $x_i, y_i$ run through distinct square classes represented by $x_i^2 - y_i^2 w$ in $F^*/F^{**}$.

Thus for the case of $G_{F(\sqrt{w})}$ where $F$ is a finite field we have two cases:

(1) $(-1)$ is a square in $F$. Then, letting $x_i = 0$, $y_i = 1$ $x_i^2 - y_i^2 w = -w = w$ in $F^*/F^{**}$, so that we may choose $\sqrt{w} = g$ the non-square class in $F(\sqrt{w})$.

(2) $(-1)$ is not a square in $F$. In order to find $g$ the non-square class in $F(\sqrt{w}) = F(\sqrt{-1})$, we must solve the equation $x^2 + y^2 = -1$. Then $g = x + y\sqrt{-1}$ is the new square class. We could as well choose $g = 1 + x\sqrt{-1}$, since $1^2 - x^2(-1) = 1 + x^2 = -y^2 =$ non-square class in $F^*$.

Now the Witt group of any finite field is generated by the 1 - dimensional forms <1>, <g>, where $g$ is a non-square. We have just computed $g$ for a quadratic extension. We now compute $tr_*$ in terms of the $g$ given in cases 1,2 above.

(1) $(-1)$ is a square in $F$. $g = \sqrt{w}$. $W(F(\sqrt{w}))$ is generated by the 1 - dimensional forms <1> and <√w>. We compute.

$$tr_*\langle 1\rangle = \begin{matrix} & 1 & \sqrt{w} \\ 1 & 2 & 0 \\ \sqrt{w} & 0 & 2w \end{matrix}$$

Here the matrix of $tr_*\langle 1\rangle$ is with respect to the basis $1, \sqrt{w}$ as indicated. It has discriminant $-4w$ which is non-trivial.

$$\text{tr}_* \langle\sqrt{w}\rangle = \begin{array}{cc} & \begin{array}{cc} 1 & \sqrt{w} \end{array} \\ \begin{array}{c} 1 \\ \sqrt{w} \end{array} & \begin{pmatrix} 0 & 2w \\ 2w & 0 \end{pmatrix} \end{array}$$

In this case the discriminant $4w^2$ is trivial.

By additivity, $\text{tr}_* \langle 1\rangle \oplus \langle\sqrt{w}\rangle \neq 0$, and case (2) follows.

(2) $(-1)$ is not a square. $g = 1 + x\sqrt{-1}$.

$$\text{tr}_* \langle 1\rangle = \begin{array}{cc} & \begin{array}{cc} 1 & \sqrt{-1} \end{array} \\ \begin{array}{c} 1 \\ \sqrt{-1} \end{array} & \begin{pmatrix} 2 & 0 \\ 0 & -2 \end{pmatrix} \end{array}$$

The discriminant 4 is trivial.

$$\text{tr}_* \langle 1+x\sqrt{-1}\rangle = \begin{array}{cc} & \begin{array}{cc} 1 & \sqrt{-1} \end{array} \\ \begin{array}{c} 1 \\ \sqrt{-1} \end{array} & \begin{pmatrix} 2 & -2x \\ -2x & -2 \end{pmatrix} \end{array}$$

The discriminant
$4(1 + x^2) = 4(-y^2)$
$\quad\quad\quad = $ non-square
$\quad\quad\quad = $ non-trivial.

Again, by additivity $\text{tr}_*\langle 1\rangle \oplus \langle 1 + x\sqrt{-1}\rangle \neq 0$, and $\text{tr}_*$ is an isomorphism on the fundamental ideal, with kernel $C_2$ as claimed.

(3) Finally, let $-$ be non-trivial on $E$, and trivial on $K$. Let $F$ be the fixed field of $-$. Then $E \supset F \supset K$.

$\text{tr}_*: H(E) \to W(K)$ is the composition

$$H(E) \xrightarrow{\text{tr}_*} W(F) \xrightarrow{\text{tr}_*} W(K)$$

By Jacobson's Theorem [M,H 115], we have the exact sequence $0 \to H(E) \xrightarrow{\text{tr}_*} W(F)$, and $\text{tr}_*$ is injective into the fundamental ideal in $W(F)$. By part (2) of this theorem, $W(F) \xrightarrow{\text{tr}_*} W(K)$, $\text{tr}_*$ is an isomorphism on the fundamental ideal in W(F). Thus, $tr_*$ is 1 - 1, $tr_*: H(E) \to W(K)$.

For $p = 2$, the fundamental ideal is trivial, so that $tr_* = 0$ as claimed. □

4. The conductor and $T(M)$

In order to finish describing $f_3$, the last task is to describe which - invariant maximal ideals $M$ in S have - invariant maximal ideals in D lying over them. This is related to the conductor.

**Definition 4.1** The conductor of D over S is the largest set C which is an ideal in both D and S.

We shall need the following theorem.

**Theorem 4.2** If A factors as $A = \prod_{i=1}^{n} P_i$ where A is prime to the conductor C, then

$$A \cap S = \prod_{i=1}^{n} (P_i \cap S).$$

Proof: See [G 38].

We shall also need a few results from ideal theory.

**Lemma 4.3** Let D be a domain, A,B,C ideals in D. Let C be generated by k elements, $C = \langle c_1, \ldots c_k \rangle$. Then $AC = BC$ implies $A^k \subset B$, and similarly $B^k \subset A$.

Proof: It clearly suffices to show an arbitrary product $(a_1 \ldots a_k)$ of $k$ elements in $A$ is contained in $B$. $A^k$ is generated by finite sums of such products.

Since $AC = BC$ we may write

$$a_1c_1 = \sum_j b_{j1}c_j, \ldots, \quad a_kc_k = \sum_j b_{jk}c_j$$

This system of equations can then be written as

$$0 = -a_ic_i + b_{1i}c_1 + b_{2i}c_2 + \ldots + b_{ki}c_k, \quad i = 1, \ldots, k.$$

Solving for $c_i$, using Cramer's rule we obtain $\Delta c_i = 0$ where $\Delta$ = determinant of coefficient matrix. $c_i \neq 0$ yields $\Delta = 0$. However, the determinant $\Delta$ can be written as $\pm(a_1 \ldots a_k) + b$ where $b \in B$. Thus $(a_1 \ldots a_k) \in B$ as desired. □

Lemma 4.4 *If a prime ideal* $P$ *in* $D$ *factors as* $P = A \cdot B$, *then* $A = D$ *or* $B = D$ (*D a Dedekind Domain*) (A Dedekind Domain has unique factorization).

Proof: $P = AB$ clearly implies $P \supset A$ or $P \supset B$. Say $P \supset A$. Then we may write $A = PW$, $W$ an ideal in $D$. Hence $P = P(WB) = PQ$. So $P \cdot D = P \cdot Q$. Since $D$ is Dedekind, $P$ is generated by 2 elements, [O'M], and we may apply Lemma 4.3.

$D^2 = D \subset Q$. Hence $Q = D = WB$, so that $W = B = D$, and $P = A$. □

Note: This Lemma is also clearly true for an order $S$.

We also recall the following:

*Theorem 4.5 Let* $S \subseteq D$ *be rings, with* $D$ *integral over* $S$. *Let* $M$ *be a prime ideal in* $S$. *Then there exists a prime ideal* $P$ *in* $D$ *with* $P \cap S = M$.

*Proof*: See [A,M 62]. □

With these preliminaries, we are in a position to give a sufficient condition for a - invariant maximal ideal in $S$ to have a unique - invariant maximal ideal in $D$ lying over it.

*Theorem 4.6 Let* $M$ *be a - invariant maximal in* $S$. *If* $M$ *is prime to the conductor* $C$, *then* $D$ *has a unique - invariant maximal ideal* $P$ *with* $P \cap S = M$.

*Proof*: Let $M$ be maximal in $S$. By Theorem 4.6, we can find $P$ maximal in $D$ with $P \cap S = M$. We claim that $P$ is unique, and hence - invariant, for clearly $\overline{P} \cap S = \overline{M} = M$. So suppose $P_i \neq P$ with $P_i \cap S = M$. $P_i$ may be $\overline{P}$, or some other maximal ideal in $D$. Each such $P_i$ will clearly appear in the factorization of $DM = \prod_{i=1}^{w} P_i$. We claim that each $P_i \cap S = M$. Clearly, $P_i \cap S \supseteq M$, since $P_i \supseteq DM$. However, $P_i \cap S \neq M$ implies that $P_i \cap S = S$, so that $1 \in P_i$, contradiction. Thus $P_i \cap S = M$.

Now note that $DM \cap S = M$. This follows since $DM \cap S \supseteq M$; $DM \subseteq P_i$ and $P_i \cap S = M$ so $DM \cap S \subseteq P_i \cap S = M$.

We now apply Theorem 4.2. Since $DM$ is prime to $C$,

$$DM \cap S = M = \prod_{i=1}^{w} (P_i \cap S) = \prod_{i=1}^{w} M.$$

However, for $M$ a prime ideal, by Lemma 4.4, $w = 1$, and consequently $P$ is unique. □

For an order $S \subsetneq D$, the conductor $C \neq 0$. For let $m \neq 0 \in Z$ satisfy $mD \subsetneq S$. Then $(m)$ generates an ideal in $S$ which is also an ideal in $D$.

Thus, we see that the cardinality of the set

$$T(M) = \{P: P \cap S = M, \quad P = \bar{P}\}$$

is one except possibly for those ideals $M$ which are not prime to $C$. Since $C \neq 0$, there are only a finite number of maximal ideals $M$ in $S$ which are not prime to $C$. Theorems 3.1 and 4.6 then enable us to compute the map $f_3$. The set $T(M)$ must be computed explicitly only at the finite set of ideals not prime to $C$.

Given $P$ in $D$, we should like to relate $D/P$ to $S/P \quad S$.

__Definition 4.7__ _Let_ $P$ _be a maximal ideal in_ $D$, _so that_ $P \quad S = M$ _is a maximal ideal in_ $S$. [A,Mc 61] _We will say_ $S$ _is integrally closed at_ $M$ _if_ $S(M)$, _the localization of_ $S$ _at_ $M$, _is integrally closed._

$D$ is integrally closed, hence so also is $D(P)$. [A,Mc 62] Thus, $S$ is integrally closed at $M$ if and only if $D(P) = S(M)$.

__Proposition 4.8__ _If_ $D(P) = S(M)$, _then_ $D/P = S/M$.

_Proof:_ $D/P = D(P)/m(P) = S(M)/m(M) = S/M$. □

Note that $D$ and $S$ have the same quotient field, $E$, and $D(P) = S(M)$ except at finitely many primes. It follows by 4.8 then that $D/P = S/M$ with only finitely many exceptions also.

# Chapter VIII THE GLOBAL BOUNDARY

Section 1 describes the coupling effect due to Stoltzfus [Sf] between various $\partial(D)$.

In Section 2 we prove $\partial\colon \mathcal{W}(k,Q) \to W(k,Q/Z)$ is onto when $k = \pm 1$. This is crucial to studying the octagon over $Z$.

## 1. <u>The coupling invariants</u>

We recall our notation: $W(k,K;f)$ denotes Witt equivalence classes of triples $[M,B,\ell]$ where the characteristic polynomial of $\ell$ is a power of the $T_k$ fixed irreducible polynomial $f(x)$. By taking anisotropic representatives, Proposition IV 1.11 identified this as:

$$W(k,K;F) \cong W(k,K;S) \quad \text{where} \quad S = Z[t]/(f(t)).$$

We used the notation $\partial(S)$ to denote the restriction of the boundary map to $W(k,K;S)$. In this section we wish to emphasize the polynomial $f$ rather than the module structure. Thus we use the notation of $\partial(f)$ to denote the restriction of the boundary map to $W(k,K;f)$. Of course $\partial(f)$ is really $\partial(S)$.

We have the commutative diagram

$$\begin{array}{ccccccc} W(k,Z) & \xrightarrow{i} & \mathcal{W}(k,Q) & \to & \displaystyle\bigoplus_{f \in \beta} W(k,Q;f) & \xrightarrow{q_f} & W(k,Q;f) \\ & & & & & & \downarrow \partial(f) \\ & & & \searrow^{\varepsilon(f)} & & & W(k,Q/Z;f) \end{array}$$

We label the composition $\partial(f) \circ q_f \circ i \equiv \varepsilon(f)$.

It follows that there is the commutative diagram:

$$\begin{array}{ccc} W(k,Z) & \xrightarrow{\oplus\, \varepsilon(f)} & \bigoplus_f W(k,Q/Z;f) \\ \downarrow i & & \downarrow \alpha_1 \\ W(k,Q) & \xrightarrow{\partial} & W(k,Q/Z) \end{array}$$

The map $\alpha_1$ just adds up all the terms in $W(k,Q/Z)$. We wish to measure how the various groups $W(k,Q/Z;f)$ couple together.

__Theorem 1.1__ __There is an exact sequence__

$$0 \to \bigoplus_{f \in \beta} W(k,Z;f) \xrightarrow{i} W(k,Z) \xrightarrow{\oplus\, \varepsilon(f)} \bigoplus_{f \in \beta} W(k,Q/Z;f)$$

$$\xrightarrow{\alpha} W(k,Q/Z) \bigoplus_{f \in \beta} \text{coker } \partial(f).$$

__Comment__: $\alpha$ is onto provided $\partial: W(k,Q) \to W(k,Q/Z)$ is onto. This is the topic of Section 7.

__Proof__: The map $\alpha$ is $\alpha_1$ on the first factor, and the appropriate projection into the cokernel on the other factor. $i$ just adds up terms in $W(k,Z)$, as does $\alpha_1$.

We begin by considering the commutative diagram:

$$\begin{array}{ccccc}
 & 0 & & 0 \\
 & \downarrow & & \downarrow \\
 & \bigoplus_f W(k,Z;f) & \xrightarrow{i} & W(k,Z) \\
 & \downarrow j_1 & & \downarrow \\
0 \to & \bigoplus_f W(k,Q;f) & \xrightarrow{i_1} & W(k,Q)
\end{array}$$

$i$ is then clearly 1 - 1 since all the other maps are.

To check exactness at $W(k,Z)$, suppose $i\langle \bigoplus_f L_f, B_f, t_f\rangle = \langle L,B,t\rangle$. Then $q_f(L \otimes Q)$ contains the self dual lattice $L_f = L_f^{\#}$ hence $\varepsilon(f)(L) = 0$. Thus $\varepsilon(f) \circ i \equiv 0$.

Conversely, suppose $(\bigoplus \varepsilon(f)) \langle L,B,t\rangle \equiv 0$. Let $M = L \otimes Q = \bigoplus_f M(f)$. Each $M(f)$ has $\varepsilon(f)(M(f)) \sim 0$, so let $N(f)$ be a metabolizer for $\varepsilon(f)(M(f))$. Exactly as in the proof of exactness for $\partial$, this yields a self-dual lattice and hence an element in $W(k,Z;f)$; which under $i$ is mapped to $M(f)$. Thus, ker $(\bigoplus \varepsilon(f)) \subseteq$ im $i$.

Next we show $\alpha \circ (\bigoplus \varepsilon(f)) = 0$. For the first factor, this follows by the commutative diagram $\alpha_1 \circ (\bigoplus \varepsilon(f)) = \partial \circ i = 0$. The other components are also 0 since $\varepsilon(f) = \partial(f) \circ q_f \circ i$ so that elements in the image of $(\bigoplus \varepsilon(f))$ are in the image of $\partial(f)$, thus 0 in coker $\partial(f)$.

Conversely, suppose we are given a collection $\bigoplus_f [M(f),B(f),t(f)]$ of torsion forms in $\bigoplus_f W(k,Q/Z;f)$, which are in the kernel of $\alpha$. Write these as $\bigoplus [M_f,B,t]$. Then each $[M_f,B,t]$ is the trivial element in cokernel of $\partial(f)$, hence $[M_f,B,t]$ is in the image of $\partial(f)$.

So let $\partial(f)[M_f',B',t'] = [M_f,B,t]$. Then

$$\bigoplus_f [M_f,B,t] = \bigoplus_f \partial(f)[M_f',B',t'].$$

Applying $\alpha_1$, ie. adding up in $W(k,Q/Z)$, we obtain:

$$\alpha_1(\bigoplus_f [M_f,B,t]) = \partial(\bigoplus_f [M_f',B',t'])$$

However, we are assuming $\alpha_1(\bigoplus_f [M_f,B,t]) = 0$. Thus $\partial(\bigoplus_f [M_f',B',t']) = 0$. By the exactness of the boundary sequence, Theorem VI 1.5, this implies there exists $[M,B,t] \in W(k,Z)$ with $i[M,B,t] = \bigoplus_f [M_f',B',t']$. Hence

$$\begin{aligned} \oplus\partial(f)q_f i([M,B,t]) &= \oplus\partial(f)[M_f',B',t'], \\ &= \bigoplus_f [M_f,B,t] \\ &= \bigoplus_f \varepsilon(f)[M,B,t] \end{aligned}$$

so that $\ker\alpha \subset \mathrm{im}(\bigoplus_f \varepsilon(f))$. $\square$

## 2. The boundary is onto

In this Section, we derive the results needed to study the octagon over $Z$. In particular, we will show that $\partial: W(k,Q) \to W(k,Q/Z)$ is onto when $k = \pm 1$, or $k$ = positive prime, $k \equiv 2,3(4)$, or $k \equiv 1(8)$.

We also show that $\partial: A(Q) \to A(Q/Z)$ is onto, and compute the cokernel of $\partial: W^{-1}(k,Q) \to W^{-1}(k,Q/Z)$ when $k = \pm 1$.

To begin with, we recall our computations:

$$W(k,Q) \simeq \begin{cases} \bigoplus_{\substack{f \in \beta \\ \text{of type } 1}} H(Q(\Theta)) \oplus W(Q(\sqrt{k})) & k \neq 1 \\ \oplus H(Q(\Theta)) \oplus W(Q) + W(Q) & k = 1 \end{cases}$$

Also, we had:

$$W(k,Q/Z) \simeq \bigoplus_{p \nmid k} W(k,F_p) \bigoplus_{p|k} W(F_p)$$

$$\simeq [\bigoplus_{p \nmid k} \bigoplus_{\substack{f \in \beta \\ \text{of type 1}}} [H(F_p(\Theta)) \bigoplus_{k \notin F_p^{**}} W(F_p(\sqrt{k})]$$

$$\bigoplus_{\substack{k \in F_p^{**} \\ p \neq 2}} W(F_p) \oplus W(F_p)] \oplus W(F_2) \bigoplus_{\substack{p|k \\ p \neq 2}} W(F_p).$$

In this decomposition, we sum over all $T_k$ fixed irreducible polynomials $f(x)$ of type 1, where $f$ has coefficients in $F_p$. We observe that any such $f$ can be lifted (not uniquely) to a $T_k$ fixed integral polynomial. To see this write

$$f(x) = \sum_{i=0}^{2n} \bar{a}_i x^i \qquad \text{where } \bar{a}_i \in F_p.$$

Lift the first $n$ coefficients $\bar{a}_i$ to $a_i \in Z$ with $a_i \equiv \bar{a}_i \pmod p$. By Lemma III 1.4, a polynomial $g(x)$ is $T_k$ fixed if and only if its coefficients satisfy $a_i k^i = a_0 a_{2n-i}$. We define $g(x)$ by

$$g(x) = \sum_{i=0}^{2n} c_i x^i \qquad \text{where } c_i = a_i \quad i = 0, \ldots, n$$

$$c_i = a_0 a_{2n-i} k^{-i} \quad i = n+1, \ldots, 2n$$

By Lemma III 1.4, $g(x)$ is $T_k$ fixed, and clearly the mod $p$ reduction of $g(x)$ is $f(x)$, since $f(x)$ is $T_k$ fixed also.

In fact, we should observe that if $Q(\Theta)$ has a non-trivial involution $\overline{\Theta} = k\Theta^{-1}$, then the irreducible polynomial of $\Theta$ over $Q$ is $T_k$ fixed. For if

$$p(x) = \sum_{i=0}^{2n} a_i x^i \text{ is the monic irreducible polynomial of } \Theta,$$

then $p(\Theta) = p(\overline{\Theta}) = 0$. Thus $p(k\Theta^{-1}) = \sum_{i=0}^{2n} a_i(k\Theta^{-1})^i = 0$.

Multiplying by $\Theta^{2n}$, $\sum_{i=0}^{2n} a_i(k)^i \Theta^{2n-i} = 0$. Thus

$$\Theta^{2n} = -\sum_{i=1}^{2n} \frac{a_i k^i \Theta^{2n-i}}{a_0}.$$

However, $\Theta^{2n} = -\sum_{i=0}^{2n-1} a_i \Theta^i$ since $p(\Theta) = 0$.

Using the fact that $1, \Theta, \ldots \Theta^{2n-1}$ is a basis for $Q(\Theta)$ over $Q$ since $p(x)$ is irreducible, we may equate coefficients of the two sums for $\Theta^{2n}$, and obtain

$$\frac{a_i k^i}{a_0} = a_{2n-i}$$

so that $p(x)$ is $T_k$ fixed by Lemma III 1.4.

This remark is important when we show the Hermitian elements in $W(k,Q/Z)$ are in the image of boundary. Given $H(F_p(\Theta_1))$, we construct an extension $Q(\Theta)$ of $Q$, with a non-trivial involution $\overline{\Theta} = k\Theta^{-1}$, with the property that the mod p reduction of $\overline{\Theta}$ is $\Theta_1$. By the above it follows that the irreducible polynomial of $\Theta$ is $T_k$ fixed. Hence $H(Q(\Theta))$ occurs in the decomposition of $W(k,Q)$. Applying $\partial$ to this $H(Q(\Theta))$ we show that $H(F_p(\Theta_1))$ is the image of $\partial$, possibly together with some Witt contribution. However, all Witt elements are first shown to be in the image of $\partial$, so that $\partial$ is onto.

We begin our study of $\partial: W(k,Q) \to W(k,Q/Z)$ by studying $\partial|W(Q(\sqrt{k}))$. Letting $S = Z[t]/(t^2 - k)$, we previously used the notation $\partial(S)$ for $\partial|W(Q(\sqrt{k}))$.

Let $D$ denote the ring of integers in $Q(\sqrt{k})$. By [S 35] $D = S$ for $k = \pm 1$ or $p$, where $p$ is a prime $p \equiv 2$ or $p \equiv 3 \pmod 4$. For $p \equiv 1 \pmod 4$. $D = \{$all elements of the form $\frac{1}{2}(u + v\sqrt{k})$, where $u,v \in Z$ with $u \equiv v \pmod 2\}$.

We now recall the computation for $\partial(D)$ given in [M,H 94]. There is an exact sequence:

$$W(Q(\sqrt{k})) \overset{\partial}{\to} \bigoplus_{P \text{ max. in } D} W(D/P) \overset{\phi}{\to} C/C^2 \to 0.$$

where $C$ = ideal class group, and $\phi$ is defined on each generator $\langle\bar{u}\rangle$ of $W(D/P)$ by: $\langle\bar{u}\rangle \to$ ideal class of $P$ modulo $C^2$.

We must be careful. This boundary sequence is for $I = D$. We are interested in the case $I = \Delta^{-1}(D/Z)$ of the inverse different. Fortunately, in our case, $\Delta^{-1}$ and $D$ are principal orders, and we may write $\Delta^{-1} = D/\alpha$, for some $\alpha \in D$. We are of course in the special case of a quadratic extension of $Q$.

We denote $\partial'(D)$ = boundary for $I = \Delta^{-1}$; $\partial(D)$ = boundary for $I = D$ in the next Lemma.

<u>Lemma 2.1</u> <u>Scaling by</u> $\alpha$ <u>induces a commutative diagram.</u>

$$\begin{array}{ccc} W(Q(\sqrt{k})) & \xrightarrow{\partial(D)} & W(E/D) \\ \simeq\downarrow \frac{1}{\alpha} & & \simeq\downarrow \frac{1}{\alpha} \\ W(Q(\sqrt{k})) & \xrightarrow{\partial'(D)} & W(E/I) \end{array}$$

Proof: $\alpha\Delta^{-1} = D$, so that commutativity follows by definition of $\partial$.□

Thus, once we have computed $\partial(D)$, we will also have a computation for $\partial'(D)$.

To begin with we will show $C/C^2$ is trivial in the stated case. This requires some number theory; we refer to [B,S]. We shall show that for $k = \pm 1$, $p$ ; $p$ prime, $p \equiv 3\ (4)$, $\partial'(D) = \partial(D) = \partial(S)$ is onto.

We also show $\partial(S)$ is onto when $k \equiv 1\ (8)$, and compute the cokernel, a $C_2$ arising from $W(F_2)$, when $k \equiv 5\ (8)$. Caution: This Witt piece, $W(F_2)$, arising in $W(k,Q/Z)$, $k \equiv 5\ (8)$, is thus not in the image of the Witt piece $W(Q(\sqrt{k}))$ in $\mathcal{W}(k,Q)$. However, we have not shown that this $W(F_2)$ is not the image of a Hermitian piece in $\mathcal{W}(k,Q)$. This question is still open.

We now aim to prove:

Theorem 2.2 $C/C^2$ is trivial for $C$ the ideal class group in $Q(\sqrt{k})$, provided $k = \pm 1$, or $k$ a positive prime.

To set our notation, there are three classes:

(1) $k \notin F_p^{**}$ (p) remains prime in D. $e = 1$ $f = 2$

$D/P = F_p(\sqrt{k})$ where $P \cap Z = (p)$.

(2) $k \in F_p^{**}$ (p) splits in D. $e = 1$ $f = 1$

$D/P_1 = D/P_2$ where $pD = P_1 P_2$.

(3) p divides k, written $p|k$

(p) ramifies $e = 2$ $f = 1$

$D/P = F_p$ where $P \cap Z = (p)$.

We follow Borevich-Shafarevich [B,S] in defining:

__Definition 2.3__ __Two ideals A and B of D are strictly equivalent if there exists $\alpha \neq 0$ in $Q(\sqrt{k})$ satisfying $N_{Q(\sqrt{k})/Q}(\alpha) > 0$ and $A = B(\alpha)$.__

For $k < 0$, $N_{Q(\sqrt{k})/Q}(\alpha) > 0$ always, so that this is the usual definition of equivalence in the ideal class group $C$. However, if $k < 0$ and $N_{Q(\sqrt{k})/Q}(\varepsilon) = +1$ for all units $\varepsilon$, then each ideal class in $C$ will split into two classes equivalent in the strict sense. [B,S 239].

[B,S] calls A,B divisors. For the case of the maximal order D, [B,S 215], divisors correspond in a 1 - 1 fashion with ideals in D.

Notation: If $P$ is an ideal, let $[P]$ denote its equivalence class, $\langle P\rangle$ denotes its strict equivalence class. When all units $\varepsilon$ have positive norm and $k > 0$ we can write $[P] = \langle P\rangle \cup \langle\sqrt{k}P\rangle$.

__Lemma 2.4__ __$[P]$ is a square in $C$ if and only if there exists an ideal $Q \in [P]$ with $\langle Q\rangle$ strictly a square.__

__Proof__: Sufficiency is clear, since if $\langle Q\rangle$ is strictly equivalent to $\langle R^2\rangle$, then $[Q] = [R^2]$ also.

Conversely, let $[P] \in C^2$. Then there exists an ideal $Q$ with $P = \alpha Q^2$. If $N_{Q(\sqrt{k})/Q}(\alpha) > 0$, $\langle P\rangle \sim \langle Q^2\rangle$ and we are done. Otherwise, suppose $N(\alpha) = \alpha\bar{\alpha} < 0$. Consider $\bar{\alpha}P \in [P]$. $\bar{\alpha}P = \bar{\alpha}\alpha Q^2$, and $N(\alpha\bar{\alpha}) > 0$. Thus $\langle \bar{\alpha}P\rangle \sim \langle Q^2\rangle$. □

Thus to check if $[P]$ is a square, we need only check if either of its strict equivalence classes is a square.

By [B,S 246], Theorem 7, $\langle A\rangle \sim \langle B^2\rangle$ if and only if $(\frac{N'(A),\bar{D}}{p}) = +1$ for all $p|\bar{D}$. Here $N'(A)$ is the norm of $A$ [B,S 124,219]. $\bar{D}$ is the discriminant of $Q(\sqrt{k})$ over $Q$, and $(\frac{N'(A),\bar{D}}{p}) = (N'(A),\bar{D})_p$ is the Hilbert symbol.

Note: We shall use our usual notation in this section for the Hilbert symbol rather than following [B,S]. $N'(A)$ is a positive integer, see [B,S 124].

Remark: $(N'(A),\bar{D})_p = +1$ automatically for $p \nmid \bar{D}$, and $p = \infty$, [B,S 242].

We are now ready to examine $C/C^2$ for $k = \pm 1$, k a positive prime.

<u>Case I.</u> For $k = \pm 1, 2$, $D$ is a principal ideal domain, so that $C = 0 = C/C^2$.

<u>Case II.</u> Let $p > 0$ be a prime $p \equiv 1 \pmod 4$.

<u>Claim</u>: For $Q(\sqrt{p})$, $C/C^2$ is trivial.

<u>Proof</u>: In this case, equivalence coincides with strict equivalence. If $P$ is a prime ideal, we shall show $\langle P\rangle \sim \langle B^2\rangle$ by computing $(N'(P),\bar{D})_p$, as $p$ divides $\bar{D}$ = discriminant = p.

Let $P \cap Z = (q)$.

Case 1. (q) is inert. Then $N'(\mathcal{P}) = q^2$. $(q^2,p)_p = +1$.

Case 2. (q) splits, q odd, so $N'(\mathcal{P}) = q$ and $(\frac{p}{q}) = +1$.

Note: $(\frac{p}{q})$ is the Legendre symbol.

$$
\begin{aligned}
(N'(\mathcal{P}),p)_p &= (q,p)_p = (\tfrac{q}{p}) \quad \text{(so by Quadratic Reciprocity)}\\
&= (\tfrac{p}{q})(-1)^{(p-1)/2 \cdot (q-1)/2}\\
&= (\tfrac{p}{q})\\
&= +1
\end{aligned}
$$

Case 3. (q) ramifies. Again $N'(\mathcal{P}) = q = p$.

$$(p,p)_p = (p,-p)_p(p,-1)_p = (\tfrac{-1}{p}) = (-1)^{(p-1)/2} = +1$$

Case 4. (q) = (2)

(a) $p \equiv 5 \pmod 8$. (2) is inert. $N'(\mathcal{P}) = 2^2$, and we are done as in Case 1.

(b) $p \equiv 1 \pmod 8$. (2) splits.

$$(2,p)_p = (\tfrac{2}{p}) = (-1)^{(p^2-1)/8} = +1$$

Thus, by Theorem 7 from [B,S], all prime ideals $\mathcal{P}$ in $C$ are squares and $C = C^2$. □

<u>Case III.</u> Let $p > 0$ be prime, $p \equiv 3 \pmod 4$.

<u>Claim</u>: $C/C^2$ is trivial for $Q(\sqrt{p})$.

In this case, each ideal class $[\mathcal{P}]$ in $C$ splits into two strict equivalence classes. We may represent these as $\langle\mathcal{P}\rangle$ and $\langle\sqrt{p}\mathcal{P}\rangle$, since $N(\sqrt{p}) = -p < 0$.

Let $P$ be a prime ideal in $Q(\sqrt{p})$. $P \cap Z = (q)$. In this case case the discriminant $\bar{D} = 4p$. Again, we have 4 cases.

Case 1. $(q)$ is inert. $N'(P) = q^2$, and $(q^2,4p)_p = +1$ as before.

Case 2. $q$ is odd. $(q)$ splits. So $(\frac{p}{q}) = +1$. $N(P) = q$.

We compute $(N'(P),4p)_{p_i}$ for $p_i = 2$ or $p$.

$$(q,4p)_p = (\frac{q}{p}) = (\frac{p}{q})(-1)^{(p-1)/2 \cdot (q-1)/2} = (-1)^{(q-1)/2}$$

$$(q,4p)_2 = (q,p)_2 = (-1)^{(p-1)/2 \cdot (q-1)/2} = (-1)^{(q-1)/2}$$

(see [O'M 206].

If $(-1)^{(q-1)/2} = -1$, $P$ is a strict square.

If $(-1)^{(q-1)/2} = -1$, consider $\sqrt{p}P \in [P]$.

In this case, namely $(-1)^{(q-1)/2} = -1$, we will show $\langle\sqrt{p}P\rangle$ is a strict square. Hence, $[P]$ is a square in $C$.

To begin with, $N'(\sqrt{p}P) = p \cdot q$. We compute,

$$\begin{aligned}
(N'(\sqrt{p}P),4p)_p &= (pq,p)_p = (p,p)_p(q,p)_p \\
&= (p,-p)_p(p,-1)_p(-1)^{(q-1)/2} \\
&= (-1)^{(p-1)/2}(-1)^{(q-1)/2} = (-1)(-1) = +1 \\
(N(\sqrt{p}P),4p)_2 &= (pq,p)_2 = (p,p)_2(q,p)_2 \\
&= (-1)^{(p-1)/2 \cdot (p-1)/2}(-1)^{(q-1)/2 \cdot (p-1)/2} \\
&= (-1)(-1) = +1
\end{aligned}$$

Thus, again by Theorem 7, we conclude $\langle\sqrt{p}P\rangle$ is a strict square.

Case 3. $P \cap Z = (q) = (p)$, so that $(q)$ ramifies. $[P]$ contains $\sqrt{p}P$. $N'(\sqrt{p}P) = pp = p^2$. Hence $[P]$ contains a strict square class, namely $\langle\sqrt{p}P\rangle$.

Case 4. $P \cap Z = (2)$. $N'(P) = 2$. If $p \equiv 3 \pmod 8$, consider $\sqrt{p}P \in [P]$. $N'(\sqrt{p}P) = 2p$.

$$\begin{aligned}(2p,4p)_p &= (2,p)_p(p,p)_p \\ &= (-1)^{(p^2-1)/8}\left(\frac{-1}{p}\right) \\ &= (-1)(-1) = +1\end{aligned}$$

$$\begin{aligned}(2p,4p)_p &= (2p,p)_2 = (2,p)_2(p,p)_2 \\ &= (-1)^{(p^2-1)/8}(-1)^{(p-1)/2 \cdot (p-1)/2} \\ &= (-1)(-1) = +1\end{aligned}$$

Thus $[P] \in C^2$.

If $p \equiv 7 \pmod 8$, $N'(P) = 2$.

$$(2,4p)_p = (2,p)_p = \left(\frac{2}{p}\right) = (-1)^{(p^2-1)/8} = +1$$

$$(2,4p)_2 = (2,p)_2 = (-1)^{p^2-1/8} = +1$$

Again $[P] \in C^2$. We have thus completed the proof of Theorem 2.2 $\square$

<u>Remark</u>: For $k$ a negative prime congruent to 1 modulo 4, $Q(\sqrt{k})$ also has $C/C^2$ trivial. The argument is just like the above.

It is also possible for one to anaylize $Q(\sqrt{k})$, for $k = p \equiv 3 \ (4)$, in which case the above argument fails.

Corollary 2.5 For $k = \pm 1, p$ with $p \equiv 2,3 \ (4)$, $\partial'(D) = \partial(D) = \partial(S)$ is onto.

Proof: Immediate from the boundary sequence and Theorem 2.1, since $S = D$ in this case. Recall $S = Z[t]/(t^2 - k)$. □

For $p \equiv 1 \ (4)$, we apply Proposition VII 4.8. It follows that $D/P = S/P \cap S$ for $P \cap Z \neq (2)$. At (2) however, when $P \cap Z = (2)$, $D/P$ has 4 elements when $p \equiv 5 \ (8)$, for then (2) is inert and $f = 2$. For $p \equiv 1 \ (8)$, (2) splits as $P_1P_2$, $f = 1$, and $D/P_1 = D/P_2 = F_2$.

Thus, $\partial(S) = tr_* \circ \partial(P)$ cannot possibly be onto $W(F_2)$ when $p \equiv 5 \ (8)$, and is onto when $p \equiv 1 \ (8)$. We summarize,

Corollary 2.6 $\partial(S)$ is onto when $p \equiv 1 \ (8)$, and has cokernel $C_2 = W(F_2)$ when $p \equiv 5 \ (8)$. □

Thus, in order to show $\partial\colon \mathcal{W}(k,Q) \to W(k,Q/Z)$ is onto all of the Witt pieces in $W(k,Q/Z)$, it remains to hit this one last Witt piece, $W(F_2)$, when $k = p \equiv 5 \ (8)$.

We thus need to show how to find a Hermitian element in $\mathcal{W}(k,Q)$ which under $\partial$ hits $W(F_2) \in W(k,Q/Z)$ whenever $k = p \equiv 5 \ (8)$ is a positive prime. This question remains open.

Corollary 2.7 For $k = \pm 1,2,3$ (4) or $k \equiv 1$ (8), $k$ a positive prime, $\partial\colon W(k,Q) \to W(k,Q/Z)$ is onto all Witt pieces in the decomposition of $W(k,Q/Z)$.

Proof: We observe that all Witt pieces in $W(k,Q/Z)$ occur in $\bigoplus_{P \text{ max in } D} W(D/P) = W(E/D)$, $E = Q(\sqrt{k})$. By Corollaries 2.5, and 2.6 $\partial(S)$ is onto these Witt pieces. Hence, so is $\partial$ all the more so. □

Corollary 2.8 $\partial\colon A(Q) \to A(Q/Z)$ is onto all Witt pieces in $A(Q/Z)$.

Proof: Same as above, since $W(Q)$ occurs in $A(Q)$. □

Thus $\partial$ restricted to $W(Q(\sqrt{k}))$; $\partial(D) = \partial(S)$, for $S = Z[t]/(t^2 - k)$ is onto when $k$ is a positive prime, $k \equiv 1,2,3$ (4) or $k \equiv 1$ (8), or $k = \pm 1$. By onto, we mean all Witt pieces in $W(k,Q/Z)$ will be in the image of $\partial(S)$, and hence in the image of $\partial$. For these $k$, it remains to show $\partial$ is onto. To do this, we must show that all Hermitian pieces in $W(k,Q/Z)$ are in the image of $\partial$.

In this inert case, we show $\partial$ is onto by hitting each $H(F_p(\Theta_1))$ separately by $\partial$, where $\Theta_1$ satisfies a $T_k$ fixed polynomial of type 1.

First, assume $p \neq 2$; we will do the case $p = 2$ last.

Let $q$ be a prime, with $(q,p) = 1 = (2,p)$. Suppose $\Theta$ satisfies $x^2 - a_1x + k = 0$ over the fixed field of the involution $-\colon \Theta_1 \to \overline{\Theta}_1 = k\Theta_1^{-1}$. Here $a_1 \in F_p(\Theta_1 + k\Theta_1^{-1})$, the fixed field of $-$. We write $a_1 = 2b_1$, which is possible since 2 and $p$ are relatively prime.

Let $F_p(\Theta_1) = F_{p^{2n}}$ = finite field with $p^{2n}$ elements. The fixed field of $-$ is $F_{p^n}$. We shall now construct an extension $Q(\Theta)$ of $Q$, together with an involution $- : \Theta \to k\Theta^{-1}$ which is non-trivial. Further, we shall arrange that the monic irreducible polynomial of $\Theta$ when read mod $p$ is the monic irreducible polynomial of $\Theta_1$.

Let $F$ denote the fixed field of $-$ on $Q(\Theta)$.

We shall arrange for at least one prime ideal, $P$ in $O(F)$, with $P \cap Z = (q)$, to ramify in $Q(\Theta)$ over $F$. We then consider the boundary map $\partial$ restricted to $H(Q(\Theta))$, with $H(Q(\Theta))$ a direct summand of $W(k,Q)$. Since there are ramified primes, the cokernel of $\partial$ will be in terms of these. Hence, $\partial|H(Q(\Theta))$ will be onto the Hermitian piece $H(F_p(\Theta_1))$, modulo the Witt pieces in $W(k,Q/Z)$. But by 2.7, these Witt pieces have already been shown to be in the image of $\partial$. Thus, $H(F_p(\Theta_1))$ is in the image of $\partial: W(k,Q) \to W(k,Q/Z)$ as desired.

Again, $\Theta_1$ satisfies $x^2 - 2b_1x + k = 0$. We begin our construction by defining $b_2$ by the equation: $b_1{}^2 = q^2b_2 + q + k$. Suppose now that $b_2$ satisfies $x^m + c_{m-1}x^{m-1} + \ldots + c_0 = f_2(x)$ over $F_p$. We have the following field extensions:

$$F_p \subseteq F_p(b_2) \subsetneq F_p(b_1) \subsetneq F_p(\Theta_1).$$

We now choose $g_2(x) = x^m + d_{m-1}x^{m-1} + \ldots + d_0$, a monic integral irreducible polynomial, with mod $p$ reduction $f_2(x)$. We also arrange for the mod $q$ reduction of $g_2(x)$ to be irreducible. This is possible by the Chinese Remainder Theorem. Thus, both ideals $(p)$ and $(q)$ remain prime in $Q(\beta_2)$, where $\beta_2$ is a root of $g_2(x)$.

Next consider the extension of $Q(\beta_2)$ given by adjoining a root of $x^2 - (q^2\beta_2 + q + k) = 0$. Call this extension $Q(\beta_1)$. The extension $Q(\beta_2) \subsetneq Q(\beta_1)$ may or may not be proper. In any case,

<u>Lemma 2.9</u> (q) <u>does not ramify in</u> $Q(\beta_2) \subsetneq Q(\beta_1)$.

<u>Proof</u>: The different $\mathcal{D}$ of this extension is the greatest common divisor of the element differents, $(f'(\alpha))$, where $\alpha$ generates $Q(\beta_1)$ over $Q(\beta_2)$. Hence $\mathcal{D}$ divides $2\sqrt{q^2\beta_2 + q + k}$. If (q) ramifies, q divides $4(q^2\beta_2 + q + k)$, so that q divides k. This is impossible since $(q,k) = 1$. $\square$

Finally, we let $E = Q(\Theta)$, where $\Theta$ satisfies $x^2 - 2\beta_1 x + k = 0$ over $Q(\beta_1)$. Notice that the mod p reduction of $\Theta$ is $\Theta_1$. Thus we have an extension of degree 2n, $[F_p(\Theta_1): F_p]$.

There is the fixed field of the involution $\Theta \to k\Theta^{-1}$ given by $Q(\beta_1)$. We are adjoining $\sqrt{4\beta_1^2 - 4k} = 2\sqrt{\beta_1^2 - k}$ to $Q(\beta_1)$. By construction, $\beta_1^2 - k = q^2\beta_2 + q = q(q\beta_2 + 1)$. This has q-adic valuation 1, ie. (q) ramifies in $Q(\Theta)$ over Q. However, also by construction, (q) does not ramify in $Q(\beta_1)$. Hence, some prime lying over (q) must ramify in $Q(\Theta)$ over $Q(\beta_1)$.

Now consider $\partial(D): H(Q(\Theta)) \to H(E/I)$. Since there are ramified primes in $Q(\Theta)$ over the fixed field, the cokernel of $\partial(D)$ is given in terms of ramified primes. In other words, the term $H(F_p(\Theta_1))$ in H(E/I) is in the image of $\partial(D)$, modulo ramified primes. Since all the ramified primes have already been shown to be in the image of $\partial$, so also is $H(F_p(\Theta_1))$ in the image of $\partial$ as desired.

The final construction is to show that $H(F_2(\Theta_1))$ is in the image of boundary. Suppose $F_2(\Theta_1) = F_2 2n$, with fixed field $F_2 n$. Suppose $\Theta_1$ satisfies $x^2 - a_1 x + k = 0$ over $F_2 n$. Suppose $a_1$ satisfies $f_1(x)$ over $F_2$. Lift each of these polynomials to Q to obtain $Q \subsetneq Q(a) \subsetneq Q(\Theta)$.

Now consider $\partial(D): H(Q(\Theta)) \to H(E/I)$. This time the cokernel will be in terms of ramified primes, or possibly $C_2$ if there is no ramification. This does not matter, since all cokernel elements are already in the image of $\partial$ by previous work. So modulo these pieces, $\partial(D)$ hits $H(F_2(\Theta_1))$ as desired.

We have thus shown how to hit with $\partial$ a typical Hermitian term $H(F_p(\Theta_1))$ in $W(k,Q/Z)$. This of course works equally well for $\partial: A(Q) \to A(Q(Z))$. We summarize.

__Theorem 2.10__ $\partial: \mathcal{W}^1(k,Q) \to W^1(k,Q/Z)$ __is onto when__ $k = \pm 1$, __or__ $k$ __a positive prime__ $k \equiv 2,3 \ (4)$ __or__ $k \equiv 1 \ (8)$.

__Theorem 2.11__ $\partial: A(k,Q) \to A(k,Q/\mathbb{Z})$ __is onto.__

In the skew case $W^{-1}(k,Q/Z)$, we need a slight modification. The above argument does show that $\partial$ is onto all inert primes modulo ramified primes. There is in fact only one Witt piece in $W^{-1}(k,Q/Z)$, namely $W(F_2)$.

__Lemma 2.12__ $W(F_2) \subsetneq W(k,Q/Z)$ __is not in the image of__ $\partial$.

__Proof__: Consider the commutative diagram of forgetful maps.

$$\begin{array}{ccc}
\mathcal{W}^{-1}(k,Q) & \xrightarrow{\partial} & W^{-1}(k,Q/Z) \\
\downarrow f_1 & & \downarrow f_2 \\
\mathcal{W}^{-1}(Q) & \xrightarrow{\partial'} & W^{-1}(Q/Z) = W(F_2).
\end{array}$$

$f_1$ and $f_2$ are the maps which forget the degree $k$ map in the data of a degree $k$ mapping structure. $f_2$ is the identity: $W(F_2) \to W(F_2)$. However, $W^{-1}(Q) = 0$, so that $W(F_2)$ is not in the image of $\partial$.

As a consequence,

Theorem 2.13 $\partial: W^{-1}(k,Q) \to W^{-1}(k,Q/Z)$, *for* $k = \pm 1$, *or* $k =$ *prime has cokernel* $C_2$ *given by the Witt element* $W(F_2)$ *in* $W^{-1}(k,Q/Z)$ *which is not in the image of* $\partial$.

In order to understand the octagon, and apply the boundary sequences above, it is first necessary to analyze the individual maps in the octagon. We do this next, in terms of the $T_k$ fixed polynomials determining the Hermitian pieces.

## Chapter IX A DETAILED ANALYSIS OF THE OCTAGON

We have an exact octagon over a field with typical term $W^{\varepsilon}(k,F)$. We also have analyzed each term

$$W^{\varepsilon}(k,F) \simeq \oplus H^{\varepsilon}(F(\Theta)) \oplus W^{\varepsilon}(F(\sqrt{k})) \qquad k \notin F^{**}$$

$$\simeq \oplus H^{\varepsilon}(F(\Theta)) \oplus W^{\varepsilon}(F) \oplus W^{\varepsilon}(F) \quad k \in F^{**}$$

In this chapter, we analyze the maps in the octagon using this direct sum decomposition. Each of these Hermitian and Witt summands is determined by a $T_k$ fixed irreducible polynomial. In Section 1, we classify these polynomials, and discuss several cases that may arise.

In Sections 2,3,4,5 we examine the various maps in the octagon. This analysis involves determining the effect of the homomorphisms on rank mod 2, signature, and discriminant. For $F$ an algebraic number field, these invariants determine $H^{\varepsilon}(F)$ by Landherr's Theorem.

### 1. The involutions

Recall $K(F)$ = {monic polynomials, non-zero constant term, coefficients in $F$}. On $K(F)$, we have several automorphisms defined.

(1) $T_k p(t) = \dfrac{t^n}{a_0}\, p(kt^{-1})$ Denote $\overline{p}(t) = T_k(p(t))$

(2) $T_{-k} p(t) = \dfrac{t^n}{a_0}\, p(-kt^{-1})$ $p^*(t) = T_{-k}(p(t))$

(3) $T_o p(t) = (-1)^n p(-t)$ $p^o(t) = T_o p(t)$

These arise from the corresponding involutions on $F[t,t^{-1}]$ given by, respectively:

(1) $t \to kt^{-1}$ $\qquad \overline{t} = kt^{-1}$

(2) $t \rightarrow -kt^{-1}$ $\qquad$ $t* = -kt^{-1}$

(3) $t \rightarrow -t$ $\qquad$ $t^{\circ} = -t$

These involutions, together with the identity, thus determine an action of the Klein 4 group, $Z/2Z \oplus Z/2Z$ on $F[t,t^{-1}]$.

We should recall the origin of these involutions. Let $[M,B,\ell] \in W(k,F)$, with $(M,B,\ell)$ anisotropic. Let $p(t)$ be the minimal polynomial of $\ell$. As we have seen in III 1.3, we may assume $p(t)$ is a $T_k$ fixed irreducible polynomial. By III 1.7, the ideal $(p(t))$ in $F[t,t^{-1}]$ is then - invariant.

We then consider $F[t,t^{-1}]/(p(t)) \simeq F(\Theta)$. $F(\Theta)$ has an involution - induced by $\bar{\Theta} = k\Theta^{-1}$, since $(p(t))$ was - invariant. Under this involution, $[M,B,\ell]$ is identified with the Hermitian inner product space $[M,B']$ over $F(\Theta)$, with $t_* \circ B' = B$, where $t_* = \text{trace } F(\Theta)/F$. We will describe the effect of the maps in the octagon on these Hermitian inner product spaces, and the associated polynomials.

To begin with, we need a criteria to determine whether $F(\Theta)$ equals $F(\Theta^2)$, when $F(\Theta)$ has an involution induced by $T_k$ for some $k$. The answer is given by considering the following cases. We will assume that the characteristic of $F$ is not 2, since $F(\Theta) = F(\Theta^2)$ when $F$ has characteristic 2.

Notation: For the remainder of this chapter, we let

$p(t)$ = irreducible polynomial of $\Theta$ over $F$

$q(t)$ = irreducible polynomial of $\Theta^2$ over $F$.

Case 1: $F(\Theta) \neq F(\Theta^2)$ when $\bar{p}(t) = p(t)$ and $p*(t) = p(t)$.

<u>Proof</u>: In this case, the ideal $(p(t))$ is both $-$ and $*$ invariant. Hence $F[t,t^{-1}]/(p(t)) \cong F(\Theta)$ has the induced involutions: $\overline{\Theta} = k\Theta^{-1}$, $\Theta^* = -k\Theta^{-1}$, $\Theta^o = (\overline{\Theta})^* = -\Theta$. These are Galois automorphisms of $F(\Theta)$. The fixed field of $o$ is $F(\Theta^2)$. As long as $o$ is non-trivial, which happens provided the characteristic of $F$ is not 2, $F(\Theta^2) \neq F(\Theta)$ by Galois theory. □

<u>Remark</u>: Since $F(\Theta^2) \neq F(\Theta)$ in this case, we have degree $p(t) = 2$ degree $q(t)$, and consequently $p(t) = q(t^2)$.

<u>Case 2</u>: $F(\Theta) = F(\Theta^2)$ when $\overline{p}(t) = p(t)$ and $p^*(t) \neq p(t)$.

In this case, the $o$ involution is not present. Recall our notation; $q(t)$ is the minimal polynomial satisfied by $\Theta^2$ over $F$. $q(t^2)$ has $\Theta$ and $-\Theta$ as roots. Note that $q(t^2)$ has only even degree terms. The hypothesis $\overline{p}(t) = p(t)$ and $p^*(t) \neq p(t)$ implies $p(t)$ has odd degree terms. Thus, $q(t^2)$ is not irreducible. Hence, we may write

$$\text{(a)} \quad q(t^2) = p(t)p(-t)w(t).$$

This follows since $p(t) \neq p(-t)$, else $p^*(t) = (\overline{p}(-t)) = \overline{p}(t) = p(t)$, contradiction. Further, degree $q(t^2) = 2$ degree $q(t) \leq 2$ degree $p(t)$. However, by (a), degree $q(t^2) \geq 2$ degree $p(t)$. Thus, degree $q(t^2) = 2$ degree $p(t)$, and $w(t) = 1$, so $q(t^2) = p(t)p(-t)$. Hence, degree $q(t) =$ degree $p(t)$, and $F(\Theta) = F(\Theta^2)$. □

<u>Case 3</u>: $F(\Theta) = F(\Theta^2)$ when $p^*(t) = p(t)$ and $\overline{p}(t) \neq p(t)$.

Proof: Exactly as above. □

## 2. The map $I_\varepsilon: W^\varepsilon(k^2,F) \to W^\varepsilon(-k,F)$

We begin with the induction map $I_\varepsilon$ defined by

$$I_\varepsilon: [M,B,\ell] \to [M \oplus M, B \oplus -kB, \tilde{\ell}]$$

where $\tilde{\ell}(x,y) = (\ell y, x)$.

Consider $[M,B'] = [M,[,]] \in H^\varepsilon(F(\sigma))$, where $H^\varepsilon(F(\sigma))$ embeds into $W^\varepsilon(k^2,F)$ via $t_*$. To keep our notation consistent with Section 1, we write $\sigma = \Theta^2$, and $H^\varepsilon(F(\sigma)) = H^\varepsilon(F(\Theta^2))$. Here we have $F[t,t^{-1}]/(q(t)) = F(\Theta^2)$.

Case 1: $F(\Theta) \neq F(\Theta^2)$

$F(\Theta^2)$ has involution given from $T_{k^2}$, $\Theta^2 \to k^2\Theta^{-2}$. This extends in two ways to $F(\Theta)$, namely $\Theta \to k\bar{\Theta}^{-1} = \bar{\Theta}$, and $\Theta \to -k\Theta^{-1} = \Theta^*$. In fact, this is the way this case is recognized, namely both the $-$ and $*$ involutions are present on $F(\Theta)$.

We identify $[M,B']$ with $[M,B,\ell] \in W^\varepsilon(k^2,F)$. $B(x,y) = t_* \circ B'(x,y)$ and $\ell$ is multiplication by $\Theta^2$. Here $B' = [,]$, so that $B = t_* \circ [,]$. Mapping over with $I_\varepsilon$ we obtain $[M \oplus M, B \oplus -kB, \tilde{\ell}]$.

We wish now to identify the (Hermitian) form we obtain in $W^\varepsilon(-k,F)$. Recall that $\tilde{\ell}^2(x,y) = (\ell x, \ell y)$. Hence, $\tilde{\ell}$ acts as a square root of $\ell$. Thus, we wish to identify $[M \oplus M, B \oplus -kB, \tilde{\ell}]$ with an Hermitian form over $F(\sqrt{\sigma}) = F(\Theta)$. This is a question of how to define an $F(\Theta)$ - vector space structure on $M \oplus M$ compatible with the $F(\Theta^2)$ - vector space structure. The following can best be understood by considering $M \simeq F(\Theta^2)$, the one dimensional case. The identifications work equally well for $M$ arbitrary.

We now write $F(\Theta) \simeq F(\Theta^2) \dot{\oplus} F(\Theta^2)$. We are naturally thinking of $\Theta = \sqrt{\sigma}$ as the ordered pair $(0,1)$. Addition of ordered pairs is componentwise. Multiplication of pairs is given by $(a,b) \cdot (c,d) = (ac + bd\Theta^2, ad + bc)$.

If the dimension $[M: F(\Theta^2)] = n$, we now view $M \oplus M$ as a vector space over $F(\Theta)$, with dimension, $[M \oplus M, F(\Theta)] = n$ also. A basis for $(M \oplus M)/F(\Theta)$ is given by $\{(v_i,0)\}$ where $\{v_i\}$ is a basis of $M/F(\Theta^2)$. Scalar multiplication by $F(\Theta) = F(\Theta^2) \dot{\oplus} F(\Theta^2)$ is given by $(a,b) \cdot (v_i,0) = (av_i,bv_i)$ on basis elements by following the above. These operations extend linearly to make $M \oplus M$ into a vector space over $F(\Theta)$. We obtain the form $<,>$ in $H^{\varepsilon}(F(\Theta))$ given by:

$$<(x,y),(z,w)> = 1/2([x,z] + -k[y,w] - k\Theta^{-1}[x,w] + \Theta[y,z])$$

where $[,] = B': M \times M \to F(\Theta^2)$. One easily checks that this respects the vector space operations given by the identification $(x,y) = x + y\Theta$, with involution on $F(\Theta)$ given by $\Theta^* = -k\Theta^{-1}$. We compute $tr_* \circ <,>$ by using

$$tr_{F(\Theta)/F} = tr_{F(\Theta^2)/F} \circ tr_{F(\Theta)/F(\Theta^2)}.$$

Here $tr_*$ is the map induced by the appropriate trace, denoted $tr$. Note, that $tr_{F(\Theta)/F(\Theta^2)}(r) = 2r$ for $r \in F(\Theta^2)$ and

$$\mathrm{tr}_{F(\Theta)/F(\Theta^2)}(\Theta) = \mathrm{tr}\,\Theta^{-1} = 0.$$

It follows that $\mathrm{tr}_* \circ \langle,\rangle = B \oplus -kB$ as desired.

We see in this case, that $I_\varepsilon$ is identified with a map $I_\varepsilon : H^\varepsilon(F(\Theta^2)) \to H^\varepsilon(F(\Theta))$.

(1) $I_\varepsilon$ preserves rank.

This is clear since $[M: F(\Theta^2)] = [M \oplus M, F(\Theta)]$, ie. $\dim_{F(\Theta^2)}(M) = \dim_{F(\Theta)}(M \oplus M)$.

(2) Signatures.

If $k < 0$, $W(k,F) = \ker I_\varepsilon$ is all torsion. Hence, in order that $[M,[,]] \in \ker I_\varepsilon$ there must be no signatures in $H^\varepsilon(F(\Theta^2))$.

If $k > 0$, $I_\varepsilon[M,[,]] \in H^\varepsilon(F(\Theta))$ is in $W(-k,F)$; again this group is all torsion. So there is no signature in the image in this case.

(3) Discriminant.

Here we must be careful because the discriminant of $[M,[,]]$ is read in

$$F(\Theta^2 + k^2\Theta^{-2})/N_{F(\Theta^2)/F(\Theta^2 + k^2\Theta^{-2})}$$

where $N_{F(\Theta^2)/F(\Theta^2 + k^2\Theta^{-2})}$ denotes elements in $F(\Theta^2 + k^2\Theta^{-2})$ which are the norms of elements from $F(\Theta^2)$, whereas the discriminant of the image, $[M \oplus M, \langle,\rangle]$ is read

$$F(\Theta - k\Theta^{-1})/N_{F(\Theta)/F(\Theta - k\Theta^{-1})}.$$

These may be different groups as the example which follows will show.

To summarize, when $F(\Theta)$ is an algebraic number field,

Theorem 2.1 Let $[M,[,]] \in H^{\varepsilon}(F(\Theta^2))$, and assume $F(\Theta^2) \neq F(\Theta)$. Then $[M,[,]] \in \ker I_{\varepsilon}$ if and only if

(a) $M$ has even rank.

(b) $M$ has signature $0$ if $k < 0$.

(c) The discriminant of $M$ when read in

$$F(\Theta - k\Theta^{-1})/N_{F(\Theta)/F(\Theta - k\Theta^{-1})}$$

must be trivial.

Proof: (a) and (b) have already been discussed. To verify (c), we need to calculate the discriminant of $<,>$. If $\dim M = n$, this is exactly given by $(1/2)^n \mathrm{dis}([,])$, by the formula for $<,>$ applied to a 1 - dimensional form, and induction. Hence, (a), (b), (c) follow by Landherr's Theorem. □

An Example.

Let $\sigma = \Theta^2 = \sqrt{-1} = i$ $\quad \Theta = \sqrt{i}$

We now have the extensions:

$$\begin{array}{ccc} Q(i) & \leftarrow & Q(\sqrt{i}) \\ \downarrow & & \downarrow \\ Q & \leftarrow & Q(\sqrt{i} + 1/\sqrt{i}) \end{array}$$

This example is for $k = -1$.

The involution on $Q(i)$ is $i \to k^2/i = 1/i = -i$

The involution on $Q(\sqrt{i})$ is $\sqrt{i} \to 1/\sqrt{i} = -i\sqrt{i}$

By elementary number theory, 3 is not a norm in $Q(i)/Q$, since 3 is not the sum of 2 squares.

However, consider $(i - 1) + \sqrt{i}$ in $Q(\sqrt{i})$. We compute its norm in $Q(\sqrt{i})/Q(\sqrt{i} + 1/\sqrt{i})$, $((i - 1) + \sqrt{i})((-i - 1) + 1/\sqrt{i}) = 3$. Thus 3 becomes a norm. This leads us to consider the following example.

Let $k = -1$, and let $M$ be a 2 - dimensional vector space over $Q(i)$, with basis $\vec{e}_1, \vec{e}_2$. With respect to this basis, consider the Hermitian form over $Q(i)$ given by

$$[,] = \begin{pmatrix} 1 & 0 \\ 0 & -3 \end{pmatrix}$$

This 2 - dimensional form has signature 0, and discriminant +3, which is not a norm.

If $\vec{e}_1, \vec{e}_2$ is a basis for $M$ over $Q(i)$, $\vec{e}_1$, $i\vec{e}_1$, $\vec{e}_2$, $i\vec{e}_2$ is a basis for $M$ over $Q$. We thus identify the Hermitian form $[M,[,]]$ with the Witt class in $W(+1,Q)$ given by $[M, t_* \circ [,], \ell]$ where $\ell\vec{x} = i\vec{x}$, so that with respect to the basis given for $M/Q$, $\ell$ has matrix

$$\begin{array}{c} \\ \vec{e}_1 \\ i\vec{e}_1 \\ \vec{e}_2 \\ i\vec{e}_2 \end{array} \begin{array}{c} \begin{array}{cccc} \vec{e}_1 & i\vec{e}_1 & \vec{e}_2 & i\vec{e}_2 \end{array} \\ \begin{pmatrix} 0 & -1 & 0 & 0 \\ 1 & 0 & 0 & 0 \\ 0 & 0 & 0 & -1 \\ 0 & 0 & 1 & 0 \end{pmatrix} \end{array}$$

We write $B = t_* \circ [,]$.

Next, apply $I_\varepsilon$ to $[M,B,\ell]$ to obtain $[M \oplus M,\ B \oplus -kB, \tilde{\ell}]$. This in turn is identified with an Hermitian form over $Q(\sqrt{i})$. With respect to the basis $(e_1,0)$, $(e_2,0)$ for $M \oplus M = V$ over $Q(\sqrt{i})$, we obtain the form

$$\begin{pmatrix} 1/2 & 0 \\ 0 & -3/2 \end{pmatrix} = \langle , \rangle$$

Since this form has discriminant 3/4, which is a norm, it follows that $[V,\langle , \rangle] = 0$ in $H(Q(\sqrt{i}))$. In this manner, we see how the norm groups

$$Q(\Theta^2 + k^2\Theta^{-2})/N_{Q(\Theta^2)/Q(\Theta^2 + k^2\Theta^2)}$$

and

$$Q(\Theta - k\Theta^{-1})/N_{Q(\Theta)/Q(\Theta - k\Theta^{-1})}$$

give rise to kernel elements for $I_\varepsilon$.

<u>Case II</u>: $F(\Theta) = F(\Theta^2)$

To begin with, $F(\Theta^2)$ has the involution given from $T_k^2$, namely $\Theta^2 \to k^2\Theta^{-2}$. Call this involution ~ , so $\tilde{\Theta}^2 = k^2\Theta^{-2}$. Hence $\tilde{\Theta}^2\Theta^2 = k^2$, and $(\Theta\tilde{\Theta})^2 = k^2$. It follows that $\Theta\tilde{\Theta} = \pm k$. This gives us two subcases.

(a) $\tilde{\Theta} = +k\Theta^{-1}$

(b) $\tilde{\Theta} = -k\Theta^{-1}$

In either case, we begin with $[M,[,]] \in H(F(\Theta^2))$. We embed this into $W(k^2,F)$ via $t_*$ to obtain $[M,B,\ell]$ where $B = t_* \circ [,]$ and

$\ell(x) = \Theta^2 x$. Applying $I_\varepsilon$ we obtain $[M \oplus M, B \oplus -kB, \tilde{\ell}]$ where $\tilde{\ell}(x,y) = (\Theta^2 y, x)$.

(a) In case (a), consider $N \subset M \oplus M$ given by $N = \{(\Theta x, x) : x \in M$. $N$ is clearly $\tilde{\ell}$ invariant, with rank $N = 1/2$ rank$(M \oplus M)$. Further,

$$\begin{aligned}(B \oplus -kB)((\Theta x, x), (\Theta y, y)) &= B(\Theta x, \Theta y) - kB(x,y) \\ &= \Theta\bar{\Theta}B(x,y) - kB(x,y) \\ &= kB(x,y) - kB(x,y) \\ &= 0\end{aligned}$$

Thus $N \subset N^{\perp}$, from which it follows that $I_\varepsilon[M,B,\ell] = 0$. This completes case (a).

(b) In case (b), we consider the rank 1 case, $M = F(\Theta^2)$; the general case follows by diagonalizing.

Let $\vec{e}_1 = (1,0)$ and $\vec{e}_2 = (0,1)$ be a basis for $M \oplus M$ over $F(\Theta^2)$

$$\tilde{\ell}(\vec{e}_1) = 0 \cdot \vec{e}_1 + \vec{e}_2 \qquad \tilde{\ell}(\vec{e}_2) = \Theta^2 \vec{e}_1 + 0 \cdot \vec{e}_2$$

Thus viewed, the matrix of $\tilde{\ell}$ relative to $\vec{e}_1$, $\vec{e}_2$ is

$$\begin{array}{c|cc} & \vec{e}_1 & \vec{e}_2 \\ \hline \vec{e}_1 & 0 & \Theta^2 \\ \vec{e}_2 & 1 & 0 \end{array}$$

The characteristic polynomial of $\tilde{\ell}$ is $\det\begin{pmatrix} x & -\Theta^2 \\ -1 & x \end{pmatrix}$

$$= x^2 - \Theta^2 = (x + \Theta)(x - \Theta).$$

The eigenvalues are thus $\Theta, -\Theta$. We next compute the corresponding eigenvectors.

$$\begin{pmatrix} 0 & \Theta^2 \\ 1 & 0 \end{pmatrix}\begin{pmatrix} x_1 \\ x_2 \end{pmatrix} = \begin{pmatrix} \Theta x_1 \\ \Theta x_2 \end{pmatrix}$$

$\Theta^2 x_2 - \Theta x_1 = 0 \qquad x_1 - \Theta x_2 = 0$

Solving the eigenspace corresponding to eigenvalue $\Theta$ is generated by $\vec{f}_1 = (\Theta,1)$ namely $\{t\vec{f}_1 : t \in F(\Theta^2)\}$. Similarly, the eigenspace corresponding to $-\Theta$ is $\{t\vec{f}_2 = t(-\Theta,1) : t \in F(\Theta^2)\}$.

| Eigenvalue | Eigenvectors | | dimension |
|---|---|---|---|
| $\Theta$ | $t(\Theta,1)$ | $t \neq 0$ | 1 |
| $-\Theta$ | $t(-\Theta,1)$ | $t \neq 0$ | 1 |

Expressing each of $\vec{f}_i$ in terms of the $\vec{e}_i$, we have $\vec{f}_1 = \Theta\vec{e}_1 + \vec{e}_2$ $\vec{f}_2 = -\Theta\vec{e}_1 + \vec{e}_2$. This leads to the change of basis matrix

$$\begin{pmatrix} \Theta & -\Theta \\ 1 & 1 \end{pmatrix} = L,$$

so $L^{-1}\tilde{\ell}_{orig}\,L = \tilde{\ell}_{new}$, where $\tilde{\ell}_{new}$ is diagonalized with respect to the basis $\vec{f}_1, \vec{f}_2$. $\tilde{\ell}_{orig}$ denotes $\tilde{\ell}$ with respect to the basis $\vec{e}_1, \vec{e}_2$.

With $\tilde{\ell}_{new} = \begin{pmatrix} \Theta & 0 \\ 0 & -\Theta \end{pmatrix}$ and $\tilde{\ell} = \begin{pmatrix} 0 & \Theta^2 \\ 1 & 0 \end{pmatrix}$ we obtain

the diagram:

$$\begin{array}{ccc} M \oplus M & \xrightarrow{L} & M \oplus M \\ \Big\downarrow \tilde{\ell}_{new} & & \Big\downarrow \tilde{\ell} \\ M \oplus M & \xrightarrow{L} & M \oplus M \end{array}$$

$L: (x,y) \to (\Theta(x - y), x + y)$. The question arises: what inner product on $M \oplus M$ makes $L$ into an isometry, $L: (M \oplus M, b, \tilde{\ell}_{new}) \to (M \oplus M, B \oplus -kB, \tilde{\ell}_{orig})$. We compute:

$$\begin{aligned} b((x,y),(z,w)) &= (B \oplus -kB)(L(x,y), L(z,w)) \\ &= (B \oplus -kB)((\Theta(x - y), x + y), (\Theta(z - w), z + w)) \\ &= B(\Theta(x - y), \Theta(z - w)) + -kB(x + y, z + w) \\ &= \Theta\tilde{\Theta}B(x - y, z - w) \quad kB(x + y, z + w) \\ &= -kB(x - y, z - w) - kB(x + y, z + w) \\ &= -k[B(x,z) + B(y,w) - B(y,z) - B(x,w) \\ &\quad + B(x,z) + B(y,w) + B(y,z) + B(x,w)] \\ &= -2k[B(x,z) + B(y,w)]. \end{aligned}$$

Thus under the isometry $L$, we may view

$$(M \oplus M, B \oplus -kB, \tilde{\ell}) \cong (M \oplus M, b, \tilde{\ell}_{new}),$$

where $b((x,y),(z,w)) = -2k(B(x,z) + B(y,w))$ $\tilde{\ell}(x,y) = (\Theta x, -\Theta y)$
Thus, the image of $I_\varepsilon$ is in two distinct pieces, $H(F(\Theta)) \oplus H(F(-\Theta))$ in this case. $I_\varepsilon$ is given by:

$$[M,[\,,\,]] \xrightarrow{I_\varepsilon} [M,b_1] \oplus [M,b_2]$$

where $b_1(x,y) = b_2(x,y) = -2k[x,y]$. It follows that $I_\varepsilon[M,[\,,\,]]$ is non-zero if $[M,[\,,\,]] \neq 0$. We summarize with:

<u>Theorem 2.2</u> <u>Let</u> $[M,[\,,\,]] \in H^\varepsilon F(\Theta^2)$, <u>and assume</u> $F(\Theta^2) = F(\Theta)$. <u>There is an induced involution</u> $\sim$ <u>on</u> $F(\Theta^2)$ <u>given by</u> $\tilde{\Theta}^2 = k\tilde{\Theta}^{-2}$. <u>This gives rise to two cases.</u>

(a) $\tilde{\Theta} = k\Theta^{-1}$

(b) $\tilde{\Theta} = -k\Theta^{-1}$

<u>In case</u> (a), $[M,[\,,\,]] \in$ <u>kernel</u> $I_\varepsilon$.

<u>In case</u> (b), $I_\varepsilon : H^\varepsilon(F(\Theta^2)) \to H^\varepsilon(F(\Theta)) \oplus H^\varepsilon(F(-\Theta))$ is 1 - 1. □

Finally, we should observe that this analysis works equally well for $F$ a finite field. The only difference is that $H^\varepsilon(F(\Theta))$ is determined by rank mod 2 only.

In particular, when $F$ has characteristic 2, and $k$ and 2 are relatively prime, $W^\varepsilon(k,F_2) \approx \oplus H^\varepsilon(k,F_2(\Theta)) \oplus W^\varepsilon(F_2)$.

We examine the map $I_\varepsilon$. For all Hermitian summands $H^\varepsilon(k,F_2(\Theta))$, we are in case (a) of Theorem 2.2. Thus $I_\varepsilon(H^\varepsilon(k,F_2(\Theta)) \equiv 0$. We must also check what happens to $W^\varepsilon(F_2)$. So consider $[M,B,\ell]$, where $\ell x = x$. Applying $I_\varepsilon$, we obtain $[M \oplus M, B \oplus -kB, \tilde{\ell}]$. A metabolizer is given by $N = \{(x,x) : x \in M\}$. Thus, when $F$ has characteristic 2, $I_\varepsilon$ is identically 0.

3. <u>The map</u> $d_\varepsilon : W^\varepsilon(-k,F) \to A(F)$.

Recall that $d_\varepsilon$ is defined by: $[M,B,\ell] \to [M,\bar{B}]$ where $\bar{B}(x,y) = k^{-1}B(x,\ell y)$. From V 2.4, the symmetry operator for $\bar{B}$, satisfying $\bar{B}(x,y) = \bar{B}(y,sx)$ is $s = -\varepsilon k \ell^{-2}$.

We consider $[M,B'] = [M,[,]] \in H^\varepsilon(F(\Theta))$, where $F(\Theta) = F[t,t^{-1}]/(p(t))$ has induced involution $\Theta^* = -k\Theta^{-1}$. This embeds into $W^\varepsilon(-k,F)$ via $t_*$ and we identify $[M,B']$ with $[M,B,\ell] \in W^\varepsilon(-k,F)$ where $B = t_* \circ B'$, $\ell x = \Theta x$.

Apply $d_\varepsilon$ and obtain $[M,\bar{B}]$. We must now identify the Hermitian form we obtain.

For $A(F)$ this is done by using a scaled trace $t_1$ IV 2.5. $t_1$ depends on the scaling factor $u$ chosen where $u\bar{u}^{-1} = s = -\varepsilon k\Theta^{-2}$. $t_1: F(\Theta^2) \to F$. Recall that $u$ may be chosen as $u = s/(1 + s) = -\varepsilon k\Theta^{-2}/(1 + -\varepsilon k\Theta^{-2}) = -\varepsilon k/(\Theta^2 - \varepsilon k)$ and $t_1(x) = t(x\bar{u}^{-1})$ where $t$ is the usual trace. Since $s = -\varepsilon k\ell^{-2} = -\varepsilon k\Theta^{-2}$, we obtain a Hermitian form with values in $H^\varepsilon(F(\Theta^2))$.

<u>Case 1</u>. $F(\Theta) \neq F(\Theta^2)$

In this case, $[M,\bar{B}]$ may be identified with the Hermitian form in $H^\varepsilon(F(-\varepsilon k\Theta^{-2})) = H^\varepsilon(F(\Theta^2))$ given by:

$$\langle x,y\rangle = \operatorname{trace}_{F(\Theta)/F(\Theta^2)} k^{-1}[x,u\Theta y] \qquad [,] = B'$$

We then have
$$\begin{aligned} t_1\langle x,y\rangle &= (t \circ h)(k^{-1}[x,u\Theta y]) \\ &= t \circ k^{-1}[x,\Theta y] \\ &= k^{-1}B(x,\Theta y) \\ &= \bar{B}(x,y) \qquad \text{as desired.} \end{aligned}$$

(h is defined in IV 2.6)

We now examine the Witt invariants. We obtain a form in $H^{\varepsilon}(F(\Theta^2))$. In this case, $F(\Theta) \neq F(\Theta^2)$, so that the rank of $M$ as a vector space over $F(\Theta^2)$ is twice the rank of $M$ over $F(\Theta)$.

We describe the method for obtaining the other invariants by examining the one dimensional case. So assume $M = F(\Theta)$, and $[1,1] = d \in F(\Theta - k\Theta^{-1})$.

Then $\vec{e}_1 = 1$, $\vec{e}_2 = \Theta$ is a basis for $M$ over $F(\Theta^2)$. With respect to $\vec{e}_1, \vec{e}_2$ we examine the matrix of the Hermitian form $\langle,\rangle$. We assume $u = -\varepsilon k/(\Theta^2 - \varepsilon k)$. $\bar{u}$ is given by the involution $\Theta^* = -k\Theta^{-1}$; the $*$ involution extends the $-$ involution. Thus the matrix of $\langle,\rangle$ is:

$$\begin{array}{c} \\ 1 \\ \Theta \end{array} \begin{array}{c} \begin{array}{cc} \quad 1 & \qquad\qquad \Theta \end{array} \\ \begin{pmatrix} tr(dk^{-1}\bar{u}\Theta^*) & tr(dk^{-1}\bar{u}\Theta^{*2}) \\ tr(dk^{-1}\bar{u}\Theta\Theta^*) & tr(dk^{-1}\bar{u}\Theta\Theta^{*2}) \end{pmatrix} \end{array}$$

tr denotes trace $_{F(\Theta)/F(\Theta^2)}$.

Using this, one may determine the signature and discriminant of $\langle,\rangle$.

<u>Case 2.</u> $F(\Theta) = F(\Theta^2)$

In this case, $[M,B'] = [M,[,]]$ has the same rank as $[M,\langle,\rangle]$. Again we examine the 1 - dimensional case, $M = F(\Theta)$, where $[1,1] = d$. Then $\langle,\rangle$ is given by:

$$\begin{aligned} \langle 1,1\rangle &= dk^{-1}\bar{u}\Theta^* = dk^{-1}[-\varepsilon k\Theta^2/(k^2 - \varepsilon k\Theta^2)](-k\Theta^{-1}) \\ &= d[+\varepsilon k\Theta/(k^2 - \varepsilon k\Theta^2)]. \end{aligned}$$

Thus, the value of the discriminant depends on this factor $\varepsilon k\Theta/(k^2 - \varepsilon k\Theta^2)$.

When $k > 0$, $W^{\varepsilon}(-k,F)$ is all torsion, and there are no signatures. When $k < 0$, we must check that the resulting form $[M,<,>]$ has 0 signature in order that $[M,[,]]$ be in the kernel of $d_{\varepsilon}$.

Again, we examine $W^{\varepsilon}(-k,F_2)$. We must be in case (2). Rank is the only invariant, so that $d_{\varepsilon}$ is 1 - 1. It is also clear that the Hermitian forms, $H^{\varepsilon}(F_2(\Theta))$, in $W^{\varepsilon}(-k,F_2)$ are mapped under $d_{\varepsilon}$ to Hermitian forms. The form $[M,B,\ell]$, with $M = F_2$, $B = \langle 1\rangle$, $\ell$ = identity, corresponding to $W^{\varepsilon}(F_2)$, likewise maps under $d_{\varepsilon}$ to $W(F_2)$. These remarks are needed for the computation of the exact octagon over $Z$ to be made later.

## 4. The map $S_{\varepsilon}: W^{\varepsilon}(k,F) \to W^{\varepsilon}(k^2,F)$

Recall that $S_{\varepsilon}$ is defined by $[M,B,\ell] \to [M,B,\ell^2]$. Let $[M,B'] = [M,[,]] \in H^{\varepsilon}(F(\Theta))$, where $F(\Theta) = F[t,t^{-1}]/(p(t))$ has induced involution $\overline{\Theta} = k\Theta^{-1}$. Embed $[M,B']$ into $W^{\varepsilon}(k,F)$ via $t_*$ and identify $[M,B']$ with $[M,B,\ell]$, where $B = t_* \circ B'$, $\ell x = \Theta x$.

We apply $S_{\varepsilon}$ and obtain $[M,B,\ell^2]$. We wish to identify the Hermitian form we obtain.

**Case 1.** $F(\Theta) \neq F(\Theta^2)$

We clearly obtain the Hermitian form in $H^{\varepsilon}(F(\Theta^2))$ given by $[M,B_1]$, where $B_1 = tr_{F(\Theta)/F(\Theta^2)} \circ B'$. The rank of $M$ over $F(\Theta^2)$ is twice the rank of $M$ over $F(\Theta)$.

In order to examine the other invariants, consider the 1 - dimensional case, $M = F(\Theta)$. A basis for $M$ over $F(\Theta^2)$ is $\vec{e}_1 = 1$, $\vec{e}_2 = \Theta$. Suppose $B' = [,]$ has $[1,1] = d \in F(\Theta + k\Theta^{-1})$. Then with respect to the basis $1, \Theta$ $B_1$ has matrix:

$$\begin{array}{cc} & \begin{array}{cc} 1 & \Theta \end{array} \\ \begin{array}{c} 1 \\ \Theta \end{array} & \begin{pmatrix} tr(d) & tr(\bar{\Theta}d) \\ tr(\Theta d) & tr(kd) \end{pmatrix} \end{array} \qquad \bar{\Theta} = k\Theta^{-1}$$

This is with $tr$ denoting $\text{trace}_{F(\Theta)/F(\Theta^2)}$. Again, this matrix enables one to compute the signature and discriminant invariants.

Case 2. $F(\Theta) = F(\Theta^2)$

In this case $F(\Theta)$ has involution $\Theta \to \bar{\Theta} = k\Theta^{-1}$, so that $\Theta^2 \to \bar{\Theta}^2 = k^2\Theta^{-2}$.

It follows that $S_\varepsilon[M,B,\ell]$ may be identified with the Hermitian form in $H^\varepsilon(F(\Theta^2))$ given by $[M,B']$. In this case $S_\varepsilon$ is then clearly $1 - 1$.

We remark that when the characteristic of $F$ is 2, $S_\varepsilon$ is $1 - 1$ by case (2). In particular, $S_\varepsilon: W(F_2) \to W(F_2)$; where the non-trivial form $<1>$ in $W(F_2)$ is identified with the form $[M,B,\ell]$ in $W(k,F_2)$ given by: $M = F_2$, $B = <1>$, $\ell$ = identity.

5. The map $m_\varepsilon: A(F) \to W^\varepsilon(k,F)$

$m_\varepsilon$ is defined by: $[M,B] \to [M \oplus M, B_\varepsilon, \ell_\varepsilon]$, where

$B_\varepsilon((x,y),(z,w)) = B(x,w) + \varepsilon B(z,y)$

$\ell_\varepsilon(x,y) = (\varepsilon k s^{-1} y, x)$.

Let $[M,B'] = [M,[,]] \in H^\varepsilon(F(\Theta))$, where $F(\Theta) = F[t,t^{-1}]/(p(t))$ has the involution - induced by $\Theta \to \bar{\Theta} = \Theta^{-1}$. We identify $[M,B']$ with $[M,B] \in A(F)$ using a scaled trace, $t_1$. Here the symmetry operator $s$ acts as $\Theta$, and $t_1(x) = t(x\bar{u}^{-1})$, where $u\bar{u}^{-1} = \Theta$. As observed before IV 2.6, we may choose $u = \Theta/(1+\Theta)$, so that $\bar{u}^{-1} = 1 + \Theta$. Then $B = t_1 \circ B'$.

The analysis of $m_\varepsilon$ is then similar to $I_\varepsilon$. The image of $[M,[,]]$ under $I_\varepsilon$ can be viewed in $H^\varepsilon(F(\sqrt{\varepsilon k\theta^{-1}}))$. This is because $\ell_\varepsilon^2 = \varepsilon k\Theta^{-1}$.

There are two cases.

<u>Case 1.</u> $F(\Theta) \neq F(\sqrt{\varepsilon k\Theta^{-1}})$

<u>Case 2.</u> $F(\Theta) = F(\sqrt{\varepsilon k\Theta^{-1}})$

Note: Let $\alpha = \sqrt{\varepsilon k\Theta^{-1}}$. The involution on $F(\alpha)$ is then $\alpha \to k\alpha^{-1} = \bar{\alpha}$. So $(\varepsilon k\Theta^{-1}) \to k^2(\varepsilon k\Theta^{-1})^{-1}$, and $\varepsilon k\Theta^{-1} \to \varepsilon k\Theta$, $\Theta \to \Theta^{-1}$. In other words, the involution on $F(\Theta)$ extends the - involution on $F(\Theta)$.

<u>Case 1.</u> $F(\Theta) \neq F(\sqrt{\varepsilon k\Theta^{-1}})$

In this case, the dimension $M \oplus M$ over $F(\sqrt{\varepsilon k\Theta^{-1}})$ is the same as the dimension of $M$ over $F(\Theta)$. As observed previously, we can make $M \oplus M$ into a vector space over $F(\sqrt{\varepsilon k\Theta^{-1}}) = F(\alpha)$. The idea again is to view $F(\alpha) = F(\Theta) \dot{\oplus} F(\Theta)$. We view $\alpha$ as the ordered pair $(0,1)$. Multiplication of ordered pairs is given by identifying

$$
\begin{aligned}
(a,b) &\to a + b\alpha \\
(c,d) &\to c + d\alpha \\
(a,b)\cdot(c,d) &\to ac + bd(\alpha^2) + (ad + bc)\alpha \\
&= ac + bd(\varepsilon k\Theta^{-1}) + (ad + bc)\alpha \\
&\to (ac + bd(\varepsilon k\Theta^{-1}), ad + bc).
\end{aligned}
$$

Thus, $m_\varepsilon\colon H(F(\Theta)) \to H(F(\alpha))$ in this case. If $M$ has basis $\{(v_i)\}$ over $F(\Theta)$, $M \oplus M$ has basis $\{(v_i,0)\}$ over $F(\alpha)$.

Consider the form $\langle , \rangle\colon M \oplus M \to F(\alpha)$ given by

$$\langle (x,y),(z,w)\rangle = (1/2k)\bar{u}^{-1}\alpha([x,z] + (k/\alpha)[x,w] + \alpha[y,z] + k[y,w]).$$

Here $\alpha^2 = \varepsilon k\Theta^{-1}$ and $[,]$ is the Hermitian form we began with.

We must check that $\langle , \rangle$ is $\varepsilon$ Hermitian. There are the identities:

$$
\begin{aligned}
\bar{u}^{-1} &= 1 + \Theta \\
\bar{u}^{-1}\alpha &= (1 + \Theta)\sqrt{\varepsilon k\Theta^{-1}} \\
u^{-1}\bar{\alpha} &= ((1 + \Theta)/\Theta)k\sqrt{\varepsilon k\Theta^{-1}}/(\varepsilon k\Theta^{-1}) = \varepsilon\bar{u}^{-1}\alpha.
\end{aligned}
$$

Hence:

$$\overline{\langle (z,w),(x,y)\rangle} = (1/2k)u^{-1}\bar{\alpha}(\overline{[z,x]} + k\bar{\alpha}^{-1}\overline{[z,y]} + \overline{\alpha[w,x]} + k\overline{[w,y]})$$

$$= (\varepsilon/2k)\bar{u}^{-1}\alpha([x,z] + (k/\alpha)[x,w] + \alpha[y,z] + k[y,w])$$

$$= \varepsilon<(x,y),(z,w)>.$$

- denotes the involution. $[a,b] = \overline{[b,a]}$ since $[,]$ is Hermitian.

Next we compute $tr_* \circ <,>$, where $tr_*$ is $trace_{F(\alpha)/F}$.

$$tr_{F(\alpha)/F} = tr_{F(\Theta)/F} \circ tr_{F(\alpha)/F(\Theta)}.$$

$$tr_{F(\alpha)/F(\Theta)} <(x,y), (z,w)> = (1/k)\bar{u}^{-1}k[x,w] + (1/k)\bar{u}^{-1}(\varepsilon k\Theta^{-1})[y,z]$$

Hence,

$$\begin{aligned} tr_{F(\alpha)/F} <(x,y),(z,w)> &= tr_{F(\Theta)/F}(\bar{u}^{-1}[x,w] + \bar{u}^{-1}\varepsilon\Theta^{-1}[y,z]) \\ &= t_1([x,w] + \varepsilon[\Theta^{-1}y,z]) \\ &= B(x,w) + \varepsilon B(\Theta^{-1}y,z) \\ &= B(x,w) + \varepsilon B(z,y). \end{aligned}$$

Hence, $tr_* \circ <,> = B_\varepsilon$ as desired.

We have thus identified the Hermitian form we obtain in the image of $m_\varepsilon$ in this case. Rank mod 2 is clearly preserved, and we read the discriminant and signatures from the extension $F(\alpha)/F(\alpha + k\alpha^{-1})$ in order to determine if $[M,[,]]$ is in the kernel of $m_\varepsilon$.

<u>Case 2.</u> $F(\Theta) = F(\sqrt{\varepsilon k\Theta^{-1}})$.

$F(\Theta)$ has the involution $\Theta \to \bar{\Theta} = \Theta^{-1}$. Under this involution, $(\bar{\alpha}^2) = \varepsilon k\Theta$. Also $\alpha^2 = \varepsilon k\Theta^{-1}$. Thus, $(\alpha^2)(\bar{\alpha}^2) = (\alpha\bar{\alpha})^2 = (\varepsilon k)^2 = k^2$.

Hence, $\alpha\bar{\alpha} = \pm k$. This gives two cases:

(a) $\bar{\alpha} = -k\alpha^{-1}$

(b) $\bar{\alpha} = k\alpha^{-1}$

Case (a) $\bar{\alpha} = -k\alpha^{-1}$.

Let $N = \{(\alpha x, x): x \in M\}$. $N$ is clearly $\ell_\varepsilon$ invariant, since $\ell_\varepsilon(x,y) = (\alpha^2 y, x)$. Further, $N$ is self-annihilating since:

$$\begin{aligned} B_\varepsilon((\alpha x,x),(\alpha y,y)) &= B(\alpha x,x) + \varepsilon B(\alpha y,x) \\ &= B(\alpha x,y) + B(\varepsilon x,\alpha\Theta y) \\ &= B(\alpha x,y) + B(\varepsilon\bar{\alpha}\,\bar{\Theta}x,y). \end{aligned}$$

However, $\alpha^2 = \varepsilon k\Theta^{-1}$, so $\alpha = (\varepsilon k\Theta^{-1})(\alpha^{-1})$. $\varepsilon\bar{\alpha}\,\bar{\Theta} = \varepsilon(-k\alpha^{-1})(\Theta^{-1}) = -\alpha$

Thus the above equals $0$ and $N$ is a metabolizer for $[M \oplus M, B_\varepsilon, \ell_\varepsilon]$. Thus $[M,[\,,]]$ is in the kernel of $m_\varepsilon$ in this case.

Case (b) $\bar{\alpha} = k\alpha^{-1}$

As with $I_\varepsilon$, we consider the one dimensional case, $M = F(\Theta)$. Let $\vec{e}_1 = (1,0)$, $\vec{e}_2 = (0,1)$ be a basis for $M \oplus M$ over $F(\theta) = F(\alpha)$. $\ell_\varepsilon(\vec{e}_1) = 0 \cdot \vec{e}_1 + 1 \cdot \vec{e}_2$ and $\ell_\varepsilon(\vec{e}_2) = \alpha^2\vec{e}_1 + \vec{e}_2$, so that $\ell_\varepsilon$ has matrix

$$\begin{pmatrix} 0 & \alpha^2 \\ 1 & 1 \end{pmatrix}$$

with respect to $\vec{e}_1$, $\vec{e}_2$.

We now diagonalize this matrix, and obtain the diagram:

$$\begin{array}{ccc} M \oplus M & \xrightarrow{L} & M \oplus M \\ \downarrow \begin{pmatrix} \alpha & 0 \\ 0 & -\alpha \end{pmatrix} & & \downarrow \begin{pmatrix} 0 & \alpha^2 \\ 1 & 0 \end{pmatrix} \\ M \oplus M & \xrightarrow{L} & M \oplus M \end{array}$$

where $L = \begin{pmatrix} \alpha & -\alpha \\ 1 & 1 \end{pmatrix}$ is the change of basis matrix.

We compute:

$$\begin{aligned} b((x,y),(z,w)) &= B_\varepsilon(L(x,y),L(z,w)) \\ &= B_\varepsilon((\alpha(x-y),x+y),(\alpha(z-w),z+w)) \\ &= B(\alpha(x-y),(z+w)) + \varepsilon B(\alpha(z-w),(x+y)) \\ &= B(\alpha x - \alpha y, z+w) + \varepsilon B(x+y,\Theta(\alpha)(z-w)) \\ &= B(\alpha x,z) + B(\alpha x,w) - B(\alpha y,z) - B(\alpha y,w) \\ &\quad + \varepsilon B(x,\alpha\Theta z) - \varepsilon B(x,\alpha\Theta w) + \varepsilon B(y,\alpha z\Theta) - \varepsilon B(y,\alpha w\Theta). \end{aligned}$$

However, $\alpha = (\varepsilon k\Theta^{-1})\alpha^{-1} = \varepsilon\bar{\alpha}\ \bar{\alpha}$ in this case. Thus, the above becomes

$$= 2\alpha[B(x,z) - B(y,w)].$$

We may thus view the image of $m_\varepsilon$ as in $H^\varepsilon(F(\alpha)) \oplus H^\varepsilon(F(-\alpha))$ in this case, namely

$$[M,[\,,\,]] \xrightarrow{m_\varepsilon} [M,b_1,\ell_1] \oplus [M,b_2,\ell_2],$$

where

$$\begin{aligned} b_1(x,y) &= 2B(\alpha x,y) & \ell_1(x) &= \alpha x \\ b_2(x,y) &= -2B(\alpha x,y) & \ell_2(x) &= -\alpha x. \end{aligned}$$

B is an F-valued asymmetric form. We must apply the trace lemma to identify $b_1, b_2$ with $<,>_1$, $<,>_2$ where $<,>_1$ and $<,>_2$ correspond with $b_1, b_2$ above, by $tr_* \circ <,>_1 = b_1$, $tr_* \circ <,>_2 = b_2$ Then $m_\varepsilon$: $[M,[,]] \to [M,<,>_1] \oplus [M,<,>_2]$.

$<,>_1$: $M \times M \to F(\alpha)$ is defined by $<x,y>_1 = 2\bar{\bar{u}}^{-1}[\alpha x,y]$. Similarly, $<x,y>_2 = -2\bar{\bar{u}}^{-1}[\alpha x,y]$. $<,>_1$ is $\varepsilon$ Hermitian since

$$\begin{aligned}\overline{<y,x>_1} &= 2u^{-1}\overline{[\alpha y,x]} \\ &= 2u^{-1}\bar{\alpha}\overline{[y,x]} \\ &= 2\varepsilon\bar{\bar{u}}^{-1}\alpha[x,y] \\ &= 2\varepsilon\bar{\bar{u}}^{-1}[\alpha x,y] \\ &= \varepsilon<x,y>_1\end{aligned}$$

$$\begin{aligned}tr_{F(\alpha)/F}\ <x,y>_1 &= tr_{F(\alpha)/F}(2\bar{\bar{u}}^{-1}[\alpha x,y]) \\ &= t_1(2[\alpha x,y]) \\ &= 2B(\alpha x,y) \\ &= b_1(x,y)\end{aligned}$$

Hence, $[M,<,>_1]$ is merely $[M,[,]]$ with the scaling factor $2\bar{\bar{u}}^{-1}$, so that $m_\varepsilon$ is $1 - 1$ in this case.

## Chapter X THE OCTAGON OVER Z

We recall the decompositions

$$W^{\varepsilon}(k,Q/Z) \simeq \bigoplus_{p\nmid k} W(k,F_p) \oplus \bigoplus_{p|k} W(k,F_p) \qquad A(Q/Z) \simeq \bigoplus_{p} A(F_p)$$

For $p|k$, the maps in the octagon for $W(k,F_p)$ do not make sense, as these terms $W(k,F_p)$ have $k = 0$. Therefore, in this section, we assume $k = \pm 1$. Hence, by the results for a field, we restate:

<u>Lemma 1.1</u> <u>There is an exact octagon where</u> $k = \pm 1$,

$$\begin{array}{ccccccc}
 & W(k,F_p) & \xrightarrow{S_1} & W(k^2,F_p) & \xrightarrow{I_1} & W(-k,F_p) & \\
{}^{m_1}\nearrow & & & & & & \searrow^{d_1} \\
A(F_p) & & & & & & A(F_p) \\
{}_{d_{-1}}\nwarrow & & & & & & \swarrow_{m_{-1}} \\
 & W^{-1}(-k,F_p) & \xleftarrow{I_{-1}} & W^{-1}(k^2,F_p) & \xleftarrow{S_{-1}} & W^{-1}(k,F_p) &
\end{array}$$

<u>Proof</u>: This is the octagon over the field $F_p$. Taking the direct sum over all $p$, we obtain

<u>Theorem 1.2</u> <u>For</u> $k = \pm 1$, <u>there is an exact octagon</u>:

$$
\begin{array}{ccccccccc}
 & & W^1(k,Q/Z) & \overset{S_1}{\rightarrow} & W^1(k^2,Q/Z) & \overset{I_1}{\rightarrow} & W^1(-k,Q/Z) & & \\
 & {}^{m_1}\nearrow & & & & & & \searrow^{d_1} & \\
A(Q/Z) & & & & & & & & A(Q/Z) \\
 & {}_{d_{-1}}\nwarrow & & & & & & \swarrow_{m_{-1}} & \\
 & & W^{-1}(-k,Q/Z) & \overset{I_{-1}}{\leftarrow} & W^{-1}(k^2,Q/Z) & \overset{S_{-1}}{\leftarrow} & W^{-1}(k,Q/Z) & &
\end{array}
$$

Although we have yet to prove exactness of the octagon over $Z$, the homomorphisms nonetheless are defined over $Z$. It is easy then to check that we have the commutative diagram which follows. $i$ denotes the map $\times_Z Q$, and $\partial$ denotes the appropriate boundary homomorphism.

$$
\begin{array}{ccccccccc}
 & & \vdots d_{-1} & & \vdots d_{-1} & & \vdots d_{-1} & & \\
0 & \rightarrow & A(Z) & \overset{i}{\rightarrow} & A(Q) & \overset{\partial}{\rightarrow} & A(Q/Z) & \rightarrow & 0 \\
 & & \downarrow m_1 & & \downarrow m_1 & & \downarrow m_1 & & \\
0 & \rightarrow & W^1(k,Z) & \overset{i}{\rightarrow} & W(k,Q) & \overset{\partial}{\rightarrow} & W^1(k,Q/Z) & \rightarrow & 0 \\
 & & \downarrow S_1 & & \downarrow S_1 & & \downarrow S_1 & & \\
0 & \rightarrow & W^1(k^2,Z) & \overset{i}{\rightarrow} & W^1(k^2,Q) & \overset{\partial}{\rightarrow} & W^1(k^2Q/Z) & \rightarrow & 0 \\
 & & \downarrow I_1 & & \downarrow I_1 & & \downarrow I_1 & & \\
0 & \rightarrow & W^1(-k,Z) & \overset{i}{\rightarrow} & W^1(-k,Q) & \overset{\partial}{\rightarrow} & W^1(-k,Q/Z) & \rightarrow & 0 \\
 & & \downarrow d_1 & & \downarrow d_1 & & \downarrow d_1 & & \\
0 & \rightarrow & A(Z) & \overset{i}{\rightarrow} & A(Q) & \overset{\partial}{\rightarrow} & A(Q/Z) & \rightarrow & 0 \\
 & & \downarrow m_{-1} & & \downarrow m_{-1} & & \downarrow m_{-1} & & \\
0 & \rightarrow & W^{-1}(k,Z) & \overset{i}{\rightarrow} & W^{-1}(k,Q) & \overset{\partial}{\rightarrow} & W^{-1}(k,Q/Z) & & \\
 & & \downarrow S_{-1} & & \downarrow S_{-1} & & \downarrow S_{-1} & & \\
0 & \rightarrow & W^{-1}(k^2,Z) & \overset{i}{\rightarrow} & W^{-1}(k^2,Q) & \overset{\partial}{\rightarrow} & W^{-1}(k^2,Q/Z) & & \\
 & & \downarrow I_{-1} & & \downarrow I_{-1} & & \downarrow I_{-1} & & \\
0 & \rightarrow & W^{-1}(-k,Z) & \overset{i}{\rightarrow} & W^{-1}(-k,Q) & \overset{\partial}{\rightarrow} & W^{-1}(-k,Q/Z) & & \\
 & & \downarrow d_{-1} & & \downarrow d_{-1} & & \downarrow d_{-1} & &
\end{array}
$$

The last two columns are exact, as are all rows. The problem is that the last three rows are not short exact, as there is the term $W(F_2)$ in $W^{-1}(k,Q/Z)$, not in the image of $\partial$.

We recall now the diagram chase that would prove exactness of the first column if all the rows were short exact. For simplicity, we label Witt equivalence classes now with symbols x, y, z, u, v, w.

To prove: $\ker S_1 = \operatorname{im} m_1$

Let $x \in W^1(k,Z)$ have $x \in \operatorname{im} m_1$. So we can find $y \in A(Z)$ with $m_1(y) = x$. Now $S_1 \circ m_1 \circ i(y) = 0$ by exactness of the 2nd column. Hence $i \circ S_1 \circ m_1(y) = 0$ by commutativity. So $i(S_1 \circ m_1(y)) = 0$. But $i$ is $1-1$, so $S_1 \circ m(y) = S_1(x) = 0$. Thus $\operatorname{im} m_1 \subset \ker S_1$.

Pictorially:

$$
\begin{array}{ccccc}
 & y & \xrightarrow{i} & i(y) \\
 & \downarrow m_1 & & \downarrow m_1 \\
 & x & & m_1(i(y)) \\
 & \downarrow S_1 & & \downarrow S_1 \\
0 \rightarrow & S_1(x) & \xrightarrow{i} & 0
\end{array}
$$

Conversely, let $x \in W^1(k,Z)$ have $x \in \ker S_1$. The picture below will facilitate reading the proof.

$$\begin{array}{ccccc}
 & & w & \xrightarrow{\partial} & z & \xrightarrow{?} & 0 \\
 & & & & \downarrow d_{-1} & & \\
 & & y & \xrightarrow{\partial} & \partial y & & \\
 & & \downarrow m_1 & & \downarrow m_1 & & \\
x & \xrightarrow{i} & i(x) & \xrightarrow{\partial} & 0 & & \\
\downarrow & & \downarrow S_1 & & & & \\
0 & \xrightarrow{i} & 0 & & & &
\end{array}$$

$i \circ S_1(x) = 0$, hence $S_1 \circ i(x) = 0$ by commutativity. The middle column is exact, so we can find $y \in A(Q)$ with $m_1(y) = i(x)$. Now $m_1(\partial y) = (\partial \circ m_1)(y) = (\partial \circ i)(x) = 0$. Thus, by exactness of the last column, we can find $z \in W^{-1}(-k,Q/Z)$ with $d_{-1}(z) = \partial y$.

This is the point that we need $\partial \colon W^{-1}(-k,Q) \to W^{-1}(-k,Q/Z)$ is onto. _If_ $\partial$ is onto $z$, we can find $w \in W^{-1}(-k,Q)$ with $\partial w = z$. Then consider $(y - d_{-1}w)$. $\partial(y - d_{-1}w) = \partial y - \partial y = 0$. Thus, by exactness of the row, we can find $v \in A(Z)$ with $i(v) = y - d_{-1}w$. However, $(m_1 \circ i)(v) = m_1(y) - (m_1 \circ d_{-1})(w) = m_1 y = ix$. Hence, $(i \circ m_1)(v) = i(x)$. Since $i$ is $1 - 1$, $m_1 v = x$. □

We see that the problem arises from $z \notin \operatorname{im} \partial$ going under $d_{-1}$ to $\partial y$, which is in the image of $\partial$. However, we have explicitly calculated what such $z$ must be. Namely, $z$ must arise from $W^{-1}(F_2)$.

We recall the computations given in the last chapter.

$$\begin{array}{ccc}
W^{-1}(F_2) & \xrightarrow{S_{-1}} & W^{-1}(F_2) \\
W^{-1}(F_2) & \xrightarrow{I_{-1}} & 0 \\
W^{-1}(F_2) & \xrightarrow{d_{-1}} & W^{-1}(F_2)
\end{array}$$

We may thus conclude that the octagon is exact over $Z$ with one possible exception, the term

$$A(Z) \xrightarrow{d_{-1}} W^{+1}(k,Z) \xrightarrow{S_1} W^{+1}(k^2,Z).$$

We must carefully analyze exactness at $W^{+1}(k,Z)$. To begin with, consider $W(F_2) \in A(Q/Z)$. This is the source of the problem.

Consider $[V,B] \in A(Q)$, where $V = \langle\vec{e}_1\rangle$, and $B = [\,,\,]$, with $[\vec{e}_1,\vec{e}_1] = 2$.

We apply $\partial$ to $[V,B]$. So let $L = \langle\vec{e}_1\rangle$ be a $Z$ - lattice. $L^\# = \langle(1/2)\vec{e}_1\rangle$ and $\langle(1/2)\vec{e}_1,(1/2)\vec{e}_1\rangle = 1/2$. It follows that $\partial[V,B] = W(F_2)$ (meaning $\partial([V,B]) \neq 0$ in $W(F_2)$ ).

Next, apply $m_1$ to $[V,B]$. Since $s$ = identity is the symmetry operator, we obtain: $[V \oplus V, B_\varepsilon, \ell_\varepsilon]$. $V \oplus V$ has basis $(1,0) = (\vec{e}_1,0) = \vec{f}_1$ and $(0,1) = (0,\vec{e}_1) = \vec{f}_2$. With respect of $\vec{f}_1$, $\vec{f}_2$ $B_\varepsilon$ has matrix

$$\begin{array}{cc} & \begin{array}{cc}\vec{f}_1 & \vec{f}_2\end{array} \\ \begin{array}{c}\vec{f}_1\\ \vec{f}_2\end{array} & \begin{pmatrix} 0 & 2 \\ 2\varepsilon & 0 \end{pmatrix}\end{array}$$

since $B_\varepsilon((x,y),(z,w)) = B(x,w) + \varepsilon B(z,y)$. $\ell_\varepsilon$ has matrix

$$\begin{pmatrix} 0 & \varepsilon k \\ 1 & 0 \end{pmatrix}$$

since $\ell_\varepsilon(x,y) = (\varepsilon k s^{-1} y, x)$. Of course, $\varepsilon = +1$ in this case.

Now $\partial[V \oplus V, B_\varepsilon, \ell_\varepsilon] = \partial \circ m_1[V,B] = m_1 \circ \partial[V,B] = m_1 \circ d_{-1}(W^{-1}(F_2)) = 0$. In fact, we may apply $\partial$ by:

Let $L$ be the $Z$ - lattice $\langle \vec{f}_1, \vec{f}_2 \rangle$. Then $L^\# = \langle (1/2)\vec{f}_1, (1/2)\vec{f}_2 \rangle$ A metabolizer for $L^\#/L$ is $N = \langle (1/2)\vec{f}_1 + (1/2)\vec{f}_2 \rangle$. There is the projection $L^\# \overset{q}{\to} L^\#/L$. Then $q^{-1}(N)$ has basis $\{(1/2)\vec{f}_1 + (1/2)\vec{f}_2, (1/2)\vec{f}_1 - (1/2)\vec{f}_2\}$, which we may write as $\{\vec{g}_1, \vec{g}_2\}$. This enables us to construct an element in $W^{+1}(k,Z)$, which when tensored with $Q$ yields $[V \oplus V, B_\varepsilon, \ell_\varepsilon]$. The element is $W = \langle \vec{g}_1, \vec{g}_2 \rangle$ as a $Z$ - module, with inner product

$$\begin{array}{cc} & \begin{array}{cc} \vec{g}_1 & \vec{g}_2 \end{array} \\ \begin{array}{c} \vec{g}_1 \\ \vec{g}_2 \end{array} & \begin{pmatrix} 1 & 0 \\ 0 & -1 \end{pmatrix} \end{array}$$

with respect to $\vec{g}_1, \vec{g}_2$, and degree $k$ map

$$\begin{pmatrix} (k+1)/2 & (-k+1)/2 \\ (-1+k)/2 & (-k-1)/2 \end{pmatrix}$$

with respect to $\vec{g}_1, \vec{g}_2$, where $\vec{g}_1 = (1/2)\vec{f}_1 + (1/2)\vec{f}_2$, and $\vec{g}_2 = (1/2)\vec{f}_1 - (1/2)\vec{f}_2$. This follows since $\ell_\varepsilon$ is

$$\begin{pmatrix} 0 & k \\ 1 & 0 \end{pmatrix}$$

with respect to $\vec{f}_1, \vec{f}_2$. We label this element $[W, b_1, t_1] = x$ where $W = \langle \vec{g}_1, \vec{g}_2 \rangle$,

$$b_1 = \begin{pmatrix} 1 & 0 \\ 0 & -1 \end{pmatrix}$$

and,

$$t_1 = \begin{pmatrix} (k+1)/2 & (-k+1)/2 \\ (-1+k)/2 & (-k-1)/2 \end{pmatrix}$$

We observe that $[W,b_1,t_1] = x$ has order 2. When $k = -1$, $x$ has order two since every element in $W(-1,Z)$ has order two. (IV 4) For if $[W,B,\ell] \in W(-1,Z)$, $\{(x,\ell x)\}$ will be an $\ell \oplus \ell$ invariant self-annihilating subspace of $[W \oplus W, B \oplus B, \ell \oplus \ell]$, hence $W \oplus W \sim 0$. When $k = +1$, we consider $[V \oplus V, B, \ell] = y = ix$. Since $i$ is one-to-one, $2x = 0$ if and only if $2y = 0$. With the matrices given,

$$B_\varepsilon = \begin{pmatrix} 0 & 2 \\ 2 & 0 \end{pmatrix} \qquad \ell_\varepsilon = \begin{pmatrix} 0 & 1 \\ 1 & 0 \end{pmatrix}$$

Clearly $\{(r,s,r,-s) : r,s \in V\}$ is a metabolizer for

$$(V \oplus V \oplus V \oplus V,\ B \oplus B,\ \ell \oplus \ell) = 2y,$$

so that $2y = 0 = 2x$.

<u>Lemma 1.3</u> $x$ <u>above is not in the image of</u> $m_1$, <u>but</u> $x$ <u>is in the kernel of</u> $S_1$.

<u>Proof</u>: By construction, $x$ is in the kernel of $S_1$. This follows since $i \circ S_1(x) = S_1 \circ i(x) = S_1 \circ m_1[V,B] = 0$. $i$ is $1-1$, so $S_1(x) = 0$.

The picture below explains the proof that $x$ is not in the image of $m_1$.

$$\begin{array}{ccccc}
 & & v & \to & \partial v \\
 & & & & \downarrow d_{-1} \\
z & \to & y & \xrightarrow{\partial} & \partial y \neq 0 \text{ in } W(F_2) \\
\vdots\, m_1 & & \downarrow\, m_1 & & \\
x & \xrightarrow{i} & i(x) & & \\
\downarrow\, S_1 & & & & \\
0 & & & &
\end{array}$$

Suppose $m_1(z) = x$. Then $m_1 \circ i(z) = i \circ m_1(z) = i(x)$. However, $m_1(y) = ix$ also. Thus, $m_1(y - i(z)) = 0$. By exactness of the middle column, there exists $v$ with $d_{-1}(v) = y - i(z)$. Now consider $\partial v$. $d_{-1} \circ \partial(v) = \partial \circ d_{-1}v = \partial(y - i(z)) = \partial y$. However, by construction, $\partial y \neq 0$ in $W(F_2)$.

The question then is: Can $\partial v$ have $d_{-1}(\partial v) = u \neq 0$ in $W^{-1}(F_2)$? Clearly $\partial(v) \neq u$ as $u$ is not in the image of $\partial$. However Hermitian summands are mapped under $d_{-1}$ to Hermitian summands by the results of the last chapter. Thus, no such $v$ can exist, and hence $x$ is not in the image of $m_1$. □

<u>Lemma 1.4</u> <u>If</u> $x_1 \in \ker S$, <u>then</u> <u>either</u> $x \in \operatorname{im} m_1$ <u>or</u> $x_1 - x \in \operatorname{im} m_1$, <u>where</u> $x = [W,b_1,t_1]$ <u>as</u> <u>described</u> <u>before</u> <u>Lemma 1.3</u>.

<u>Proof</u>: The picture below may be useful.

$$\begin{array}{ccccc}
 & & w & \dashrightarrow & z_1 \\
 & & & & \downarrow d_{-1} \\
 & & y_1 & \xrightarrow{\partial} & \partial y_1 \\
 & & \downarrow m_1 & & \downarrow m_1 \\
x_1 & \xrightarrow{i} & ix_1 & \xrightarrow{\partial} & 0 \\
\downarrow S_1 & & \downarrow S_1 & & \\
0 & \rightarrow & 0 & &
\end{array}$$

Since $S_1x_1 = 0$, $S_1 \circ ix_1 = 0$, and there exists $y_1$ with $m_1y_1 = ix_1$ by exactness of the middle column. Now consider $\partial y_1$. By commutativity, $m_1 \circ \partial y_1 = 0$, so that by exactness of the last column we can find $z_1$ with $d_{-1}(z_1) = \partial y_1$. The question is: Is $z_1$ in the image of $\partial$? If the answer is yes, we proceed as in the general case and conclude $x_1$ is in the image of $m_1$. If the answer is no, consider $z_1 + u$. This clearly must be in the image of $\partial$ say $\partial w = z_1 + u$ where $u \neq 0$ in $W^{-1}(F_2)$. Again we proceed as before and conclude $x_1 - x$ is in the image of $m_1$. □

We may now state the theorem we have been aiming for:

<u>Theorem 1.5</u> <u>The following octagon is exact</u>.

$$\begin{array}{ccccccc}
 & W^{+1}(k,Z) & \xrightarrow{S_1} & W^{+1}(k^2,Z) & \xrightarrow{I_1} & W^{+1}(-k,Z) & \\
\nearrow m_1 \oplus i & & & & & & \searrow d_1 \\
A(Z) \oplus C_2 & & & & & & A(Z) \\
\nwarrow d_{-1} & & & & & & \swarrow m_{-1} \\
 & W^{-1}(-k,Z) & \xleftarrow{I_{-1}} & W^{-1}(k^2,Z) & \xleftarrow{S_{-1}} & W^{-1}(k,Z) &
\end{array}$$

Proof: Here $C_2$ denotes the element $[W,B_1,t_1] = x$ constructed prior to Lemma 1.3. As we have seen the only question is exactness at $W^{+1}(k,Z)$.

Let $x_1 \in \text{im}\,(m_1 \oplus i)$, where $i$ is the identity on $x$; so $x_1 = m_1y$ or $x_1 = m_1y + x$. Applying $S_1$, we obtain

$$\begin{aligned} S_1(m_1y) + S_1x &= S_1m_1y \qquad \text{by } 1.3 \\ &= 0 \end{aligned}$$

Conversely, let $x_1 \in \ker S$. By 1.4, either $x_1 \in \text{im}\, m_1$ or $x_1 - x \in \text{im}\, m_1$. In either case, $x_1$ is in the image of $m_1 \oplus i$ as desired.

Finally, we should remark that adding the term $x$ to $A(Z)$ does not create new kernel elements for $m_1$. This is because $x \in W^{+1}(k,Z)$ is not in the image of $m_1$ by 1.3. □

Remark: The reason no problem occurred with

$$S_{-1}\colon\ W^{-1}(F_2) \to W^{-1}(F_2)$$

is that neither term is in the image of $\partial$.

## NOTATION

This is a list of commonly used symbols and abbreviations. A complete definition and description of each symbol is generally given in the text. This list is intended as an index of symbols.

| Symbol | Description |
|---|---|
| Z | The ring of integers |
| Q | The rational numbers |
| D | A Dedekind domain |
| E | The quotient field of D |
| - | An involution on E |
| F | The fixed field of - |
| $E^*$ | Units in E |
| $E^{**}$ | Squares in E |
| $NE^*$ | Norms from E |
| $F^*/NE^*$ | Group of - fixed elements modulo norms |
| O(E) | Dedekind ring of integers in E namely D |
| O(F) | Dedekind ring of integers in F |
| $O(E)^*$ | Units in O(E) |
| S | An order in D |
| $\mathcal{P}$ | A prime ideal in O(E) |
| $\mathcal{M}$ | A prime ideal in S |
| P | A prime ideal in O(F) |
| $O_E(\mathcal{P})$ | Local ring of integers at $\mathcal{P}$ |
| $\tilde{O}_E(\mathcal{P})$ | Completion of O(E) at $\mathcal{P}$ |

| Symbol | Description |
|---|---|
| I | Fractional ideal in O(E) |
| $I(\mathcal{P})$ | I localized at $\mathcal{P}$ |
| $M(\mathcal{P})$ | M localized at $\mathcal{P}$ |
| $m(\mathcal{P})$ | The localization of $\mathcal{P}$ in $O_E(\mathcal{P})$ |
| $D/\mathcal{P}$ | The residue field $O(E)/\mathcal{P}$, also isomorphic to $O_E(\mathcal{P})/m(\mathcal{P})$ |
| $\pi_{\mathcal{P}}$ or $\pi$ | Uniformizer for $\mathcal{P}$ |
| $\pi_P$ | Uniformizer for P |
| $\| \|_{\mathcal{P}}$ | $\mathcal{P}$-adic valuation on E |
| $\| \|_P$ | P-adic valuation on F |
| $v_{\mathcal{P}}$ | Additive version of $\| \|_{\mathcal{P}}$ |
| $v_P$ | Additive version of $\| \|_P$ |
| (M,B) | Inner product space |
| [M,B] | Witt equivalence class of (M,B) |
| $(M,B,\ell)$ | Degree k mapping structure |
| $\ell$ | Degree k map |
| $[M,B,\ell]$ | Witt equivalence class of $(M,B,\ell)$ |
| $\ell^*$ | Right adjoint operator of $\ell$ |
| ${}^*\ell$ | Left adjoint operator of $\ell$ |
| $Ad_R B$ | Right adjoint map of B |
| $Ad_L B$ | Left adjoint map of B |
| $N_R$ | Right orthogonal complement of N |
| $N_L$ | Left orthogonal complement of N |
| $N^{\perp}$ | Orthogonal complement of N |
| $W^{+1}(K)$ | Witt equivalence classes of inner product spaces, (M,B) with B symmetric |

| Symbol | Description |
|---|---|
| $W^{+1}(k,K)$ | Witt equivalence classes of degree k mapping structures with B symmetric |
| $A(K)$ | Witt equivalence classes of inner product spaces, (M,B) with no symmetry requirements |
| s | The symmetry operator |
| $\mathcal{W}(k,K)$ | Degree k mapping structures $(M,B,\ell)$ under Witt equivalence, with the characteristic polynomial of $\ell$ integral |
| $\mathcal{A}(K)$ | The characteristic polynomial of s is integral |
| Ann M | Annihilator of M |
| Ext | Cokernel of Hom functor |
| K(F) | Monic polynomials, coefficients in F, nonzero constant term |
| GK(F) | Grothendieck group associated to K(F) |
| $H^2(k;K(F))$ | Cohomology group $H^2(C_2;GK(F))$ |
| $\mathcal{B}$ | Basis for $H^2(k;K(F))$ as an $F_2$-vector space |
| $T_k$ | Involution on K(F) |
| $\mathcal{D}(E/F) = \Delta(E/F)$ | Different of E over F |
| $\Delta^{-1}(E/F)$ | Inverse different of E over F |
| $N_{E/F}$ | The norm map of E over F |
| C | Ideal class group |
| char | Characteristic |
| deg | Degree |
| dim | Dimension |
| det | Determinant |
| Dis | Discriminant |

| Symbol | Description |
|---|---|
| rk | Rank |
| ker | Kernel |
| im | Image |
| $e_i$ | Ramification index |
| $f_i$ | Residue field degree |
| J | Fundamental ideal of even rank forms |
| ~ | Witt equivalence relation |
| □ | End of proof |
| $p\|k$ | p divides k |
| $(p_1,p_2) = 1$ | $p_1$ and $p_2$ are relatively prime |
| t, tr, $t_*$ | Various trace maps |
| $t_1$ | Scaled trace |
| $\partial$ | Boundary homomorphism |
| $L^\#$ | Dual lattice |
| $T(M)$ | $\{P : P \cap S = M\}$ |
| sgn | Signature |
| $F_q$ | Finite field with q elements |
| D[t] | Polynomials over D |
| $D[t,t^{-1}]$ | Finite Laurant series over D |
| $C_p$ | Cyclic group with p elements |
| $(y,\sigma)_p$ | Hilbert symbol |
| $\left(\frac{p}{q}\right)$ | Legendre symbol |
| Q/Z | Quotient as a Z-module |
| W(k,F;f) | Witt equivalence classes $[M,B,\ell]$ with $f(\ell) = 0$ |

| Symbol | Description |
| --- | --- |
| $W(k,F;D)$ | Witt equivalence classes with a compatible D-module structure |
| $\partial(D)$ | $\partial$ restricted to $\dot{W}(k,F;D)$ |
| $\partial(D,P) = \partial(P)$ | The localization of $\partial(D)$ at $P$ |
| $\langle n_1,\ldots,n_t\rangle$ | The space spanned by $n_1,\ldots,n_t$ |
| $G_F$ | $F^*/F^{**}$ |
| $N'(A)$ | Norm of ideal [B,S 124] |

## REFERENCES

[A,C,H] Alexander, J.P., P.E. Conner and G.C. Hamrick, Odd Order Group Actions and Witt Classification of Inner Products, Lecture Notes in Math. 625, Springer-Verlag, Heidelberg, Germany (1977).

[A] Artin, E., Algebraic Numbers and Functions, Gordon and Breach, N.Y. (1967).

[A,Mc] Atiyah, M.F. and I.G. Macdonald, Introduction to Commutative Algebra, Addison-Wesley, Reading, Mass. (1969).

[B,S] Borevich, Z.I. and I.R. Shafarevich, Number Theory, Academic Press, N.Y. (1966).

[B-1] Bourbaki, N., Éléments 24(Algèbre 9), Formes sesquilinéaires et formes quadratiques, Hermann, Paris, France (1959).

[B-2] Bourbaki, N., Commutative Algebra, Hermann, Addison-Wesley, Reading, Mass. (1972).

[C] Conner, P.E., Notes on the Witt Classification of Hermitian Innerproduct Spaces over a Ring of Algebraic Integers, Univ. of Texas Press, (1979).

[C,E] Cartan, H. and S. Eilenberg, Homological Algebra, Princeton U. Press, Princeton, N.J. (1956).

[E,L] Elman, Richard and T.Y. Lam, Quadratic Forms Under Algebraic Extensions, Math. Ann. 219, (1976) (21-42).

[G] Gilmer, R.W. Some Relations Between Ideals in Different Integral Domains, A Thesis, L.S.U. (1960).

[G,F] Gross, H. and H.R. Fisher, Non-real Fields k and Infinite Dimensional k-Vector Spaces, Math Ann. 159 (1965), (285-308).

[G,S] Geramita, Anthony V. and Jennifer Seberry, Orthogonal Designs, Quadratic Forms and Hadamard Matrices, Lecture Notes in Pure and Applied Mathematics, Volume 45, Marcel Dekker, N.Y. (1979).

[H] Herstein, I., Topics in Algebra, Blaisdell Pub. Co., Waltham, Mass. (1974).

[H,N,K] Hirzebruch, F., W.D. Neumann and S.S. Koh, Differentiable Manifolds and Quadratic Forms, Marcel Dekker, N.Y. (1971).

[K-1] Kaplansky, I., Linear Algebra and Geometry, Chelsea Pub. Co., N.Y. (1974).

[K-2] Kaplansky, I., Commutative Rings, Allyn and Bacon, Boston, Mass. (1970).

[Kr] Kreck, M., Bordism of Diffeomorphisms, Bull, A.M.S. 82 (1976)

[Lm] Lam, T.Y., The Algebraic Theory of Quadratic Forms, W.A. Benjamin Inc. Reading, Mass. (1973).

[Lh] Landherr, W., Äquivalenz Hermitescher Formen über einen beliebigen algebraischen Zahlkörper,Abh. Math. Sem., Hamburg Univ. 11 (1936).

[L-1] Lang, S., Algebra, Addison-Wesley Pub. Co., Reading, Mass. (1970).

[L-2] Lang, S., Algebraic Number Theory, Addison-Wesley, Reading, Mass. (1970).

[M] Maclane, S., Homology, Springer-Verlag, Heidelberg, Germany (1963).

[M,H] Milnor, J. and D. Husemoller, Symmetric Bilinear Forms Springer-Verlag, Heidelberg, Germany, (1973).

[O'M] O'Meara, O.T., Introduction to Quadratic Forms, Academic Press, N.Y. (1963).

[Q] Quinn, Frank, Open Book Decompositions and the Bordism of Automorphisms, Topology, Vol. 18 No. 1, (1979) (55-73).

[R-1] Rotman, J.J. Notes on Homological Algebra, Van Nostrand Reinhold Co., N.Y. (1970).

[R-2] Rotman, J.J., The Theory of Groups, Allyn and Bacon, Boston, Mass. (1965).

[S] Samuel, P., Algebraic Theory of Numbers, Hermann-Houghton-Mifflin, Boston, Mass. (1970).

[Sf-1] Stoltzfus, N.W., Unravelling the integral knot concordance group, Mem. Amer. Math. Soc., Prov., R.I. (1977).

[Sf-2] Stoltzfus, N.W., The Algebraic Relationship Between Quinn's Invariant of Open Book Decomposition and the Isometric Structure of the Monodromy, (to appear).

[V] Vick, J., Homology Theory, Academic Press, N.Y. (1973).

[W] Warshauer, M.L., Diagonalization up to Witt, Pacific Journal of Mathematics, (to appear).

[Z,S-1] Zariski, O. and P. Samuel, Commutative Algebra, Vol. 1, Springer-Verlag, Heidelberg, Germany (1958).

[Z,S-2] Ibid. Vol. 2, (1960).

## INDEX

Vol. 759: R. L. Epstein, Degrees of Unsolvability: Structure and Theory. XIV, 216 pages. 1979.

Vol. 760: H.-O. Georgii, Canonical Gibbs Measures. VIII, 190 pages. 1979.

Vol. 761: K. Johannson, Homotopy Equivalences of 3-Manifolds with Boundaries. 2, 303 pages. 1979.

Vol. 762: D. H. Sattinger, Group Theoretic Methods in Bifurcation Theory. V, 241 pages. 1979.

Vol. 763: Algebraic Topology, Aarhus 1978. Proceedings, 1978. Edited by J. L. Dupont and H. Madsen. VI, 695 pages. 1979.

Vol. 764: B. Srinivasan, Representations of Finite Chevalley Groups. XI, 177 pages. 1979.

Vol. 765: Padé Approximation and its Applications. Proceedings, 1979. Edited by L. Wuytack. VI, 392 pages. 1979.

Vol. 766: T. tom Dieck, Transformation Groups and Representation Theory. VIII, 309 pages. 1979.

Vol. 767: M. Namba, Families of Meromorphic Functions on Compact Riemann Surfaces. XII, 284 pages. 1979.

Vol. 768: R. S. Doran and J. Wichmann, Approximate Identities and Factorization in Banach Modules. X, 305 pages. 1979.

Vol. 769: J. Flum, M. Ziegler, Topological Model Theory. X, 151 pages. 1980.

Vol. 770: Séminaire Bourbaki vol. 1978/79 Exposés 525–542. IV, 341 pages. 1980.

Vol. 771: Approximation Methods for Navier-Stokes Problems. Proceedings, 1979. Edited by R. Rautmann. XVI, 581 pages. 1980.

Vol. 772: J. P. Levine, Algebraic Structure of Knot Modules. XI, 104 pages. 1980.

Vol. 773: Numerical Analysis. Proceedings, 1979. Edited by G. A. Watson. X, 184 pages. 1980.

Vol. 774: R. Azencott, Y. Guivarc'h, R. F. Gundy, Ecole d'Eté de Probabilités de Saint-Flour VIII-1978. Edited by P. L. Hennequin. XIII, 334 pages. 1980.

Vol. 775: Geometric Methods in Mathematical Physics. Proceedings, 1979. Edited by G. Kaiser and J. E. Marsden. VII, 257 pages. 1980.

Vol. 776: B. Gross, Arithmetic on Elliptic Curves with Complex Multiplication. V, 95 pages. 1980.

Vol. 777: Séminaire sur les Singularités des Surfaces. Proceedings, 1976-1977. Edited by M. Demazure, H. Pinkham and B. Teissier. IX, 339 pages. 1980.

Vol. 778: SK1 von Schiefkörpern. Proceedings, 1976. Edited by P. Draxl and M. Kneser. II, 124 pages. 1980.

Vol. 779: Euclidean Harmonic Analysis. Proceedings, 1979. Edited by J. J. Benedetto. III, 177 pages. 1980.

Vol. 780: L. Schwartz, Semi-Martingales sur des Variétés, et Martingales Conformes sur des Variétés Analytiques Complexes. XV, 132 pages. 1980.

Vol. 781: Harmonic Analysis Iraklion 1978. Proceedings 1978. Edited by N. Petridis, S. K. Pichorides and N. Varopoulos. V, 213 pages. 1980.

Vol. 782: Bifurcation and Nonlinear Eigenvalue Problems. Proceedings, 1978. Edited by C. Bardos, J. M. Lasry and M. Schatzman. VIII, 296 pages. 1980.

Vol. 783: A. Dinghas, Wertverteilung meromorpher Funktionen in ein- und mehrfach zusammenhängenden Gebieten. Edited by R. Nevanlinna and C. Andreian Cazacu. XIII, 145 pages. 1980.

Vol. 784: Séminaire de Probabilités XIV. Proceedings, 1978/79. Edited by J. Azéma and M. Yor. VIII, 546 pages. 1980.

Vol. 785: W. M. Schmidt, Diophantine Approximation. X, 299 pages. 1980.

Vol. 786: I. J. Maddox, Infinite Matrices of Operators. V, 122 pages. 1980.

Vol. 787: Potential Theory, Copenhagen 1979. Proceedings, 1979. Edited by C. Berg, G. Forst and B. Fuglede. VIII, 319 pages. 1980.

Vol. 788: Topology Symposium, Siegen 1979. Proceedings, 1979. Edited by U. Koschorke and W. D. Neumann. VIII, 495 pages. 1980.

Vol. 789: J. E. Humphreys, Arithmetic Groups. VII, 158 pages. 1980.

Vol. 790: W. Dicks, Groups, Trees and Projective Modules. IX, 127 pages. 1980.

Vol. 791: K. W. Bauer and S. Ruscheweyh, Differential Operators for Partial Differential Equations and Function Theoretic Applications. V, 258 pages. 1980.

Vol. 792: Geometry and Differential Geometry. Proceedings, 1979. Edited by R. Artzy and I. Vaisman. VI, 443 pages. 1980.

Vol. 793: J. Renault, A Groupoid Approach to C*-Algebras. III, 160 pages. 1980.

Vol. 794: Measure Theory, Oberwolfach 1979. Proceedings 1979. Edited by D. Kölzow. XV, 573 pages. 1980.

Vol. 795: Séminaire d'Algèbre Paul Dubreil et Marie-Paule Malliavin. Proceedings 1979. Edited by M. P. Malliavin. V, 433 pages. 1980.

Vol. 796: C. Constantinescu, Duality in Measure Theory. IV, 197 pages. 1980.

Vol. 797: S. Mäki, The Determination of Units in Real Cyclic Sextic Fields. III, 198 pages. 1980.

Vol. 798: Analytic Functions, Kozubnik 1979. Proceedings. Edited by J. Ławrynowicz. X, 476 pages. 1980.

Vol. 799: Functional Differential Equations and Bifurcation. Proceedings 1979. Edited by A. F. Izé. XXII, 409 pages. 1980.

Vol. 800: M.-F. Vignéras, Arithmétique des Algèbres de Quaternions. VII, 169 pages. 1980.

Vol. 801: K. Floret, Weakly Compact Sets. VII, 123 pages. 1980.

Vol. 802: J. Bair, R. Fourneau, Etude Géometrique des Espaces Vectoriels II. VII, 283 pages. 1980.

Vol. 803: F.-Y. Maeda, Dirichlet Integrals on Harmonic Spaces. X, 180 pages. 1980.

Vol. 804: M. Matsuda, First Order Algebraic Differential Equations. VII, 111 pages. 1980.

Vol. 805: O. Kowalski, Generalized Symmetric Spaces. XII, 187 pages. 1980.

Vol. 806: Burnside Groups. Proceedings, 1977. Edited by J. L. Mennicke. V, 274 pages. 1980.

Vol. 807: Fonctions de Plusieurs Variables Complexes IV. Proceedings, 1979. Edited by F. Norguet. IX, 198 pages. 1980.

Vol. 808: G. Maury et J. Raynaud, Ordres Maximaux au Sens de K. Asano. VIII, 192 pages. 1980.

Vol. 809: I. Gumowski and Ch. Mira, Recurences and Discrete Dynamic Systems. VI, 272 pages. 1980.

Vol. 810: Geometrical Approaches to Differential Equations. Proceedings 1979. Edited by R. Martini. VII, 339 pages. 1980.

Vol. 811: D. Normann, Recursion on the Countable Functionals. VIII, 191 pages. 1980.

Vol. 812: Y. Namikawa, Toroidal Compactification of Siegel Spaces. VIII, 162 pages. 1980.

Vol. 813: A. Campillo, Algebroid Curves in Positive Characteristic. V, 168 pages. 1980.

Vol. 814: Séminaire de Théorie du Potentiel, Paris, No. 5. Proceedings. Edited by F. Hirsch et G. Mokobodzki. IV, 239 pages. 1980.

Vol. 815: P. J. Slodowy, Simple Singularities and Simple Algebraic Groups. XI, 175 pages. 1980.

Vol. 816: L. Stoica, Local Operators and Markov Processes. VIII, 104 pages. 1980.

Vol. 817: L. Gerritzen, M. van der Put, Schottky Groups and Mumford Curves. VIII, 317 pages. 1980.

Vol. 818: S. Montgomery, Fixed Rings of Finite Automorphism Groups of Associative Rings. VII, 126 pages. 1980.

Vol. 819: Global Theory of Dynamical Systems. Proceedings, 1979. Edited by Z. Nitecki and C. Robinson. IX, 499 pages. 1980.

Vol. 820: W. Abikoff, The Real Analytic Theory of Teichmüller Space. VII, 144 pages. 1980.

Vol. 821: Statistique non Paramétrique Asymptotique. Proceedings, 1979. Edited by J.-P. Raoult. VII, 175 pages. 1980.

Vol. 822: Séminaire Pierre Lelong–Henri Skoda, (Analyse) Années 1978/79. Proceedings. Edited by P. Lelong et H. Skoda. VIII, 356 pages, 1980.

Vol. 823: J. Král, Integral Operators in Potential Theory. III, 171 pages. 1980.

Vol. 824: D. Frank Hsu, Cyclic Neofields and Combinatorial Designs. VI, 230 pages. 1980.

Vol. 825: Ring Theory, Antwerp 1980. Proceedings. Edited by F. van Oystaeyen. VII, 209 pages. 1980.

Vol. 826: Ph. G. Ciarlet et P. Rabier, Les Equations de von Kármán. VI, 181 pages. 1980.

Vol. 827: Ordinary and Partial Differential Equations. Proceedings, 1978. Edited by W. N. Everitt. XVI, 271 pages. 1980.

Vol. 828: Probability Theory on Vector Spaces II. Proceedings, 1979. Edited by A. Weron. XIII, 324 pages. 1980.

Vol. 829: Combinatorial Mathematics VII. Proceedings, 1979. Edited by R. W. Robinson et al.. X, 256 pages. 1980.

Vol. 830: J. A. Green, Polynomial Representations of $GL_n$. VI, 118 pages. 1980.

Vol. 831: Representation Theory I. Proceedings, 1979. Edited by V. Dlab and P. Gabriel. XIV, 373 pages. 1980.

Vol. 832: Representation Theory II. Proceedings, 1979. Edited by V. Dlab and P. Gabriel. XIV, 673 pages. 1980.

Vol. 833: Th. Jeulin, Semi-Martingales et Grossissement d'une Filtration. IX, 142 Seiten. 1980.

Vol. 834: Model Theory of Algebra and Arithmetic. Proceedings, 1979. Edited by L. Pacholski, J. Wierzejewski, and A. J. Wilkie. VI, 410 pages. 1980.

Vol. 835: H Zieschang, E. Vogt and H.-D. Coldewey, Surfaces and Planar Discontinuous Groups. X, 334 pages. 1980.

Vol. 836: Differential Geometrical Methods in Mathematical Physics. Proceedings, 1979. Edited by P. L. García, A. Pérez-Rendón, and J. M. Souriau. XII, 538 pages. 1980.

Vol. 837: J. Meixner, F. W. Schäfke and G. Wolf, Mathieu Functions and Spheroidal Functions and their Mathematical Foundations Further Studies. VII, 126 pages. 1980.

Vol. 838: Global Differential Geometry and Global Analysis. Proceedings 1979. Edited by D. Ferus et al. XI, 299 pages. 1981.

Vol. 839: Cabal Seminar 77 – 79. Proceedings. Edited by A. S. Kechris, D. A. Martin and Y. N. Moschovakis. V, 274 pages. 1981.

Vol. 840: D. Henry, Geometric Theory of Semilinear Parabolic Equations. IV, 348 pages. 1981.

Vol. 841: A. Haraux, Nonlinear Evolution Equations- Global Behaviour of Solutions. XII, 313 pages. 1981.

Vol. 842: Séminaire Bourbaki vol. 1979/80. Exposés 543–560. IV, 317 pages. 1981.

Vol. 843: Functional Analysis, Holomorphy, and Approximation Theory. Proceedings. Edited by S. Machado. VI, 636 pages. 1981.

Vol. 844: Groupe de Brauer. Proceedings. Edited by M. Kervaire and M. Ojanguren. VII, 274 pages. 1981.

Vol. 845: A. Tannenbaum, Invariance and System Theory: Algebraic and Geometric Aspects. X, 161 pages. 1981.

Vol. 846: Ordinary and Partial Differential Equations, Proceedings. Edited by W. N. Everitt and B. D. Sleeman. XIV, 384 pages. 1981.

Vol. 847: U. Koschorke, Vector Fields and Other Vector Bundle Morphisms – A Singularity Approach. IV, 304 pages. 1981.

Vol. 848: Algebra, Carbondale 1980. Proceedings. Ed. by R. K. Amayo. VI, 298 pages. 1981.

Vol. 849: P. Major, Multiple Wiener-Itô Integrals. VII, 127 pages. 1981.

Vol. 850: Séminaire de Probabilités XV. 1979/80. Avec table générale des exposés de 1966/67 à 1978/79. Edited by J. Azéma and M. Yor. IV, 704 pages. 1981.

Vol. 851: Stochastic Integrals. Proceedings, 1980. Edited by D. Williams. IX, 540 pages. 1981.

Vol. 852: L. Schwartz, Geometry and Probability in Banach Spaces. X, 101 pages. 1981.

Vol. 853: N. Boboc, G. Bucur, A. Cornea, Order and Convexity in Potential Theory: H-Cones. IV, 286 pages. 1981.

Vol. 854: Algebraic K-Theory. Evanston 1980. Proceedings. Edited by E. M. Friedlander and M. R. Stein. V, 517 pages. 1981.

Vol. 855: Semigroups. Proceedings 1978. Edited by H. Jürgensen, M. Petrich and H. J. Weinert. V, 221 pages. 1981.

Vol. 856: R. Lascar, Propagation des Singularités des Solutions d'Equations Pseudo-Différentielles à Caractéristiques de Multiplicités Variables. VIII, 237 pages. 1981.

Vol. 857: M. Miyanishi. Non-complete Algebraic Surfaces. XVIII, 244 pages. 1981.

Vol. 858: E. A. Coddington, H. S. V. de Snoo: Regular Boundary Value Problems Associated with Pairs of Ordinary Differential Expressions. V, 225 pages. 1981.

Vol. 859: Logic Year 1979–80. Proceedings. Edited by M. Lerman, J. Schmerl and R. Soare. VIII, 326 pages. 1981.

Vol. 860: Probability in Banach Spaces III. Proceedings, 1980. Edited by A. Beck. VI, 329 pages. 1981.

Vol. 861: Analytical Methods in Probability Theory. Proceedings 1980. Edited by D. Dugué, E. Lukacs, V. K. Rohatgi. X, 183 pages. 1981.

Vol. 862: Algebraic Geometry. Proceedings 1980. Edited by A. Libgober and P. Wagreich. V, 281 pages. 1981.

Vol. 863: Processus Aléatoires à Deux Indices. Proceedings, 1980. Edited by H. Korezlioglu, G. Mazziotto and J. Szpirglas. V, 274 pages. 1981.

Vol. 864: Complex Analysis and Spectral Theory. Proceedings, 1979/80. Edited by V. P. Havin and N. K. Nikol'skii, VI, 480 pages. 1981.

Vol. 865: R. W. Bruggeman, Fourier Coefficients of Automorphic Forms. III, 201 pages. 1981.

Vol. 866: J.-M. Bismut, Mécanique Aléatoire. XVI, 563 pages. 1981.

Vol. 867: Séminaire d'Algèbre Paul Dubreil et Marie-Paule Malliavin. Proceedings, 1980. Edited by M.-P. Malliavin. V, 476 pages. 1981.

Vol. 868: Surfaces Algébriques. Proceedings 1976–78. Edited by J. Giraud, L. Illusie et M. Raynaud. V, 314 pages. 1981.

Vol. 869: A. V. Zelevinsky, Representations of Finite Classical Groups. IV, 184 pages. 1981.

Vol. 870: Shape Theory and Geometric Topology. Proceedings, 1981. Edited by S. Mardešić and J. Segal. V, 265 pages. 1981.

Vol. 871: Continuous Lattices. Proceedings, 1979. Edited by B. Banaschewski and R.-E. Hoffmann. X, 413 pages. 1981.

Vol. 872: Set Theory and Model Theory. Proceedings, 1979. Edited by R. B. Jensen and A. Prestel. V, 174 pages. 1981.